# THE SOLAR SYSTEM

*A* **SCIENTIFIC** *Book* **AMERICAN**

# THE SOLAR SYSTEM

W. H. FREEMAN AND COMPANY

San Francisco

Library of Congress Cataloging in Publication Data

Main entry under title:

The Solar system.

"A Scientific American book."
Articles from the Sept. 1975 issue of Scientific American.
Bibliography: p.
Includes index.
1. Solar system. I. Scientific American.
QB501.S62          523.2          75-28113
ISBN 0-7167-0551-6
ISBN 0-7167-0550-8 pbk.

The twelve chapters in this book originally appeared
as articles in the September 1975 issue of *Scientific
American*.

Printed in the United States of America

9 8 7 6 5 4

Cover photograph from Ames Research Center, NASA,
Moffett Field, California

# CONTENTS

**FOREWARD** *vii*

1  THE SOLAR SYSTEM, by Carl Sagan
Presenting a book about it after 18 years of the direct exploration of space.   3

2  THE ORIGIN AND EVOLUTION OF THE SOLAR SYSTEM, by A. G. W.
Cameron   The sun and the planets condensed out of a thin disk of dust and gas.   15

3  THE SUN, by E. N. Parker
The central star of the solar system and its activity are still full of surprises.   27

4  MERCURY, by Bruce C. Murray
The innermost planet is like the earth on the inside and the moon on the outside.   37

5  VENUS, by Andrew and Louise Young
Its atmosphere traps sunlight to maintain a temperature of 900 degrees Fahrenheit.   49

6  THE EARTH, by Raymond Siever
Compared with the other planets, it is marked by restless activity (including life).   59

7  THE MOON, by John A. Wood
The earth's dead satellite records events early in the history of the solar system.   69

8  MARS, by James B. Pollack
Its extraordinary surface features testify to lively activity at an earlier epoch.   81

9  JUPITER, by John H. Wolfe
More massive than all the other planets put together, it is mainly liquid hydrogen.   95

10  THE OUTER PLANETS, by Donald M. Hunten
Saturn, Uranus and Neptune somewhat resemble Jupiter. Pluto is a small maverick.   105

11  THE SMALLER BODIES OF THE SOLAR SYSTEM, by William K.
Hartmann   They range from tiny meteoroids to satellites larger than Mercury.   113

12  INTERPLANETARY PARTICLES AND FIELDS, by James A. Van Allen
The "solar wind" interacts intricately with the magnetic fields of the planets.   127

The Authors   137

Bibliographies   139

Index   141

# FOREWORD

This book presents a comprehensive picture of the new knowledge about the solar system gathered by man's first manned and unmanned expeditions beyond Earth's sheltering sky.

A picture it is, for much of the new evidence is photographic. Pictures show Mercury to be a bleak, moonlike planet, virtually without atmosphere, pitted with impact craters and hot from the too-near Sun; Venus concealed in a dense atmosphere of carbon dioxide that is stirred by titanic storms driven by the inpouring of solar energy; Mars also cratered like the moon but showing volcanic craters as well as evidence of erosion by water and by the winds of its thin atmosphere; Jupiter, the giant planet, better comprehended as a kind of cool scale model of a star that is 2.5 times brighter than the sunlight reflected from its surface, its deep atmosphere banded and mottled by the resolution of the forces of convection and rotation.

Much evidence has come in, of course, on other channels. We have learned about the elemental composition of the planets, their temperatures, and the nature of their atmospheres; about Jupiter's enormous and swiftly changing magnetic field, which nearly wrecked the approaching Pioneer 11; about the all-pervading solar wind of energetic particles racing outward from the Sun wherever the explorers have ventured. From all of this new knowledge has come deeper understanding of the formation and evolution of the Solar System. Such understanding tells us that solar systems—and so life and perhaps intelligent life—are more common in the Galaxy and the universe beyond than we had guessed before.

An important part of the story is the technology that supported the space missions. There are the extraordinary control systems that carried men on the round trip to the moon and brought Mariner 10 (Venus/Mercury) into a second and then a third productive encounter with Mercury, escorting that planet on its solar orbit. There is the equally extraordinary instrumentation that took the place of men on the planetary expeditions and returned to Earth a torrent of information.

The authors who present this picture are the vicarious spacemen whose experiments made up the payloads of these expeditions.

The chapters of this book first appeared in the September 1975 issue of SCIENTIFIC AMERICAN, the twenty-sixth in the series of single-topic issues published annually by the magazine. To our colleagues at W. H. Freeman and Company, the book-publishing affiliate of SCIENTIFIC AMERICAN, we declare herewith our appreciation for the enterprise that has made this issue so speedily available in book form.

THE EDITORS*

September 1975

*BOARD OF EDITORS: Gerard Piel (Publisher), Dennis Flanagan (Editor), Francis Bello (Associate Editor), Philip Morrison (Book Editor), Trudy E. Bell, Brian P. Hayes, Jonathan B. Piel, David Popoff, John Purcell, James T. Rogers, Armand Schwab, Jr., C. L. Stong, Joseph Wisnovsky

# THE SOLAR SYSTEM

1

PISCES

SYSTEMA TYCHONIS, *vixit circa finem Sec. XVI*

*Sic oculis,*

SYSTEMA COPE

*sic ratio*

# The Solar System

CARL SAGAN

*Presenting what is known about the sun and the bodies in orbit around it, with special reference to the knowledge gained in 18 years of exploration by space probes launched from the earth*

Imagine that the earth has been watched over the millenniums by a careful and extremely patient extraterrestrial observer. Some 4.6 billion years ago the planet completes its aggregation out of interstellar gas and dust. The last planetesimals that fall in to make the earth produce enormous impact craters; the planet heats up internally from the gravitational potential energy of its accretion and from the decay of its radioactive elements; the heavy liquid iron core separates from the lighter silicate mantle and crust; hydrogen-rich gases and condensable water are released from the interior to the surface, and a fairly straightforward organic chemistry yields complex molecules that combine into self-replicating molecular systems: the first terrestrial organisms. The rain of interplanetary boulders dwindles, and in time running water, wind, mountain building and other geological processes erase the scars of the earth's origin. A vast planetary convection engine is established that carries mantle material up through rifts in the ocean floor to form great crustal plates and then drives the material back into the mantle at the margins of the continents; collisions between plates push up chains of folded mountains, and the general configuration of land and sea, of icy and tropical regions shifts continuously. Meanwhile natural selection nominates from a wide range of alternative candidates those varieties of self-replicating molecular systems best suited to the latest change in the environment. Plants evolve that use visible light to break down water into hydrogen and oxygen, and the hydrogen escapes into space, changing the atmosphere from a reducing medium to an oxidizing one. Organisms of moderate complexity and modest intelligence eventually arise.

Throughout this sequence our imaginary observer is struck above all by the earth's isolation. Sunlight, starlight and cosmic rays, and occasionally some interplanetary debris, arrive at the earth's surface, but in all those aeons nothing save a little hydrogen and helium leaves the planet. And then, less than 20 years ago, the planet suddenly begins, like a dandelion gone to seed, to fire tiny capsules throughout the inner solar system. First they go into orbit around the earth and then to the planet's lifeless natural satellite, the moon. Six tiny capsules, larger than the rest, set down on the moon and from each two small organisms emerge, briefly explore their immediate surroundings and then sprint back to the earth, having tentatively extended a toe into the cosmic ocean. Five little spacecraft enter the hellhole of Venus' atmosphere and three of them survive some tens of minutes on the surface before being destroyed by the heat. More than a dozen

CRUCIAL DECISION in the evolution of the modern heliocentric model of planetary motion is represented symbolically in the illustration on the opposite page. The scene is a detail of a large hand-colored wood engraving of the solar system, one of a set of 30 such astronomical charts compiled by Johann Gabriel Doppelmayer of Nuremberg and published in 1742 under the title *Atlas novus coelestis*. The two circular diagrams depict the planetary system according to the two great 16th-century astronomers Tycho Brahe and Nicolaus Copernicus. The Latin inscriptions on the yellow ribbons outlining the two diagrams read in translation: "System of Tycho, who lived around the end of the 16th century" and "System of Copernicus, who lived around the beginning of the 16th century." The phrases in the italic letters just under each diagram can be translated as "Thus by eye" in the case of Tycho (who was noted primarily as an observer) and "Thus by reason" in the case of Copernicus (who was of course best known as the conceiver of the heliocentric theory). The female figure, presumably Urania, the muse of astronomy, appears to favor the Copernican system over the Tychonic system. (The Copernican system, as in the prevailing modern view, has the planets revolving around the sun at the center; the Tychonic system, following the medieval Ptolemaic tradition, continues to have the earth at the center but attempts to account for the observational data of the time by having the other planets revolve around the sun as the sun revolves around the earth.) The meaning of the symbols that appear in the chart is given in the illustration on page 6. The copy of the Doppelmayer atlas from which this reproduction was made is in the Burndy Library in Norwalk, Conn.

spacecraft are dispatched to Mars; one sends back information for a full year from its orbit around the planet. Another swings by Venus to encounter Mercury, on a trajectory that will cause it to pass close to the innermost planet many times. Two more successfully traverse the asteroid belt, fly close to Jupiter and are ejected by its gravity into interstellar space. It is clear, the observer might report, that something interesting is happening on the planet earth.

We have entered, almost without noticing it, an age of exploration and discovery unparalleled since the Renaissance, when in just 30 years European man moved across the Western ocean to bring the entire globe within his ken. Our new ocean is beyond that globe: it is the shallow disk of space occupied by the solar system. Our new worlds are the sun, the moon and the planets. In less than 20 years of space exploration we have learned more about those worlds than we have in all the preceding centuries of earthbound observation. We

are beginning to assemble that information into a new picture of our solar system.

It is useful—and somewhat humbling—to start by placing our small solar neighborhood in its proper cosmic perspective. The earth is a tiny hunk of rock and metal that rides in a flood of sunlight through the innermost recess of the solar system. Other tiny spheres of rock and metal—Mercury, Venus and Mars—move in orbit around the sun

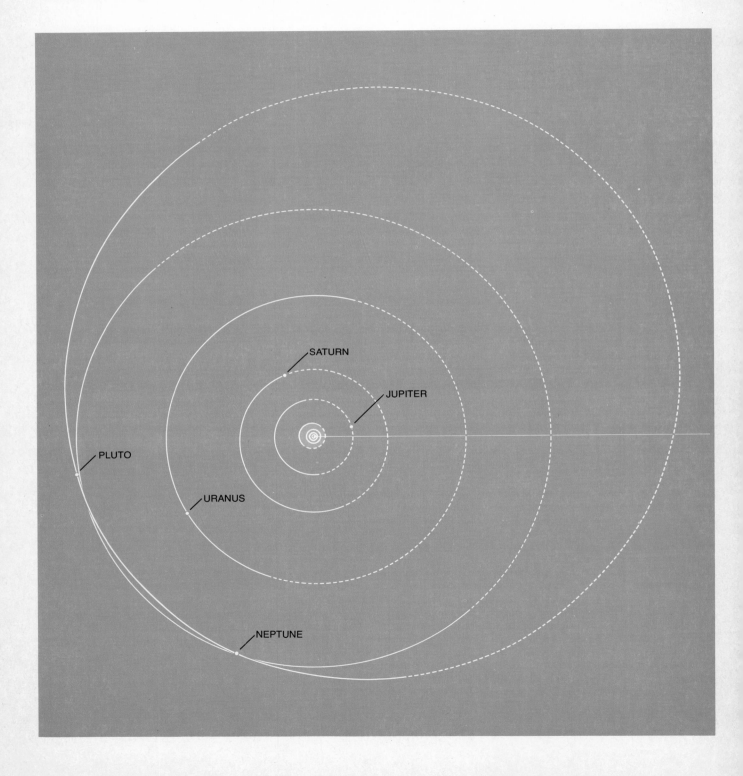

nearby. These inner planets and their satellites do not bulk very large in the solar system as a whole. Most of the mass, angular momentum and (from any extraterrestrial astronomer's viewpoint) ostensible interest of the solar system resides in the Jovian planets: four immense and rapidly rotating spheres. The inner two, Jupiter and Saturn, consist largely of hydrogen and helium; indeed, Jupiter is something like a star that failed. The outer two, Uranus and Neptune, are composed less of the lightest gases

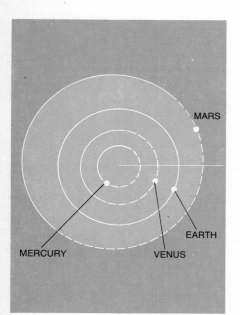

**SOLAR SYSTEM FOR SEPTEMBER is represented in the large diagram at left with the planetary orbits all drawn to the correct scale; an enlargement of the inner portion of the diagram appears above. The white dots indicate only the positions of the planets; since the mean diameter of the earth's orbit is roughly 200 times the solar diameter, even the sun would be a barely perceptible speck at the scale of either diagram. By convention the heliocentric longitude of each planet is measured in degrees of arc from the vernal equinox (*straight line at right*). The broken curves denote the portion of each planet's orbit that lies below the earth's orbital plane (the ecliptic). Pluto's orbit is anomalous in several respects. It is the only planetary orbit whose eccentricity can be distinguished from a "zero ellipse" (a circle) at the scale of such diagrams. Moreover, it has by far the greatest inclination to the ecliptic of any planetary orbit: more than 17 degrees. These and other considerations have led some to regard Pluto as being not a planet but rather an escaped satellite of Neptune. Beginning in 1987 Pluto will lose even its questionable distinction of being the outermost planet in the solar system when it slips inside the orbit of Neptune on its way to perihelion (the point on its orbit closest to the sun).**

and more of such heavier gases as methane and ammonia. Jupiter takes almost 12 years to complete its trip around the sun at a mean distance of some five astronomical units. (An astronomical unit is the mean distance of the earth from the sun, about 93 million miles or 150 million kilometers). Beyond the Jovian planets Pluto, smaller and less familiar, orbits eccentrically at about 40 astronomical units. Much farther, at about 100,000 astronomical units, are some billions of tailless comets, kilometer-size snowballs slowly circling the distant sun.

From somewhat farther away, say a few hundred thousand astronomical units, the sun would appear to the unaided eye as a bright star with no hint of its retinue of planets. That would be a distance of a few light-years (a light-year is about 60,000 astronomical units), or the characteristic separation between stars in our galaxy. From a few dozen light-years away the sun would be quite undetectable to the unaided human eye—and a distance of a few dozen light-years is only about a thousandth of the distance from the sun to the center of our galaxy. The galaxy is a vast, ponderously rotating pinwheel of some 250 billion suns, and the dense central plane of the galaxy, seen edge on, is the diffuse band across the sky that we call the Milky Way. Our galaxy is one of at least billions, and perhaps hundreds of billions, of galaxies. Our particular sun and its companion planets constitute no more than one example of a phenomenon that must surely be repeated innumerable times in the vastness of space and time.

If the 4.6 billion years of earth history were compressed into a single year, the flurry of space exploration would have begun less than a tenth of a second ago. The fundamental changes in attitude and knowledge responsible for the remarkable transformation would have filled only the past few seconds, since the first widespread application of simple lenses and mirrors for astronomical purposes in the 17th century. Before that the planets had been recognized for millenniums as being different from the "fixed stars," which appeared not to move with respect to one another. The planets (the word comes from the Greek for "wanderer") were brighter than most stars, and they moved against the stellar background. Since the sun and the moon manifestly influenced the earth, astrological doctrine held that the planets must affect human life too, but in more subtle ways. Almost none of the ancients speculated that the planets were worlds

in some sense like the earth. With the first astronomical telescope, however, Galileo was astonished and delighted to see Venus as a crescent lighted by the sun and to make out the mountains and craters of the moon. Johannes Kepler thought the craters were the constructions of intelligent beings inhabiting the moon, but Christiaan Huygens disagreed. He argued that the construction of such great circular depressions would require an unreasonably great effort—and he thought he could see natural explanations for them.

Huygens exemplified the marriage of advancing technology and experimental skills with a reasonable, skeptical mind and an openness to new ideas. He was the first to suggest that on Venus we are looking at an atmosphere and clouds, the first to understand something of the true nature of the rings of Saturn (which had seemed to Galileo like two "ears" enveloping the planet), the first to draw a picture of a recognizable marking on the surface of Mars (Syrtis Major) and the second (after Robert Hooke) to draw the Great Red Spot of Jupiter. The last two observations are of current significance because they establish the continuity of prominent planetary surface features over at least three centuries. (Huygens was, to be sure, not a thoroughly modern astronomer; he could not entirely escape the modes of belief of his time. Consider the curious argument by which he deduced the existence on Jupiter of hemp. Galileo had observed four moons traveling around Jupiter. Huygens asked a question of a kind few astronomers would ask today: Why is it that Jupiter has four moons? Well, why does the earth have one moon? Our moon's function, Huygens reasoned, apart from providing a little light at night and raising the tides, is to aid mariners in navigation. If Jupiter has four moons, there must be many mariners on that planet. Mariners imply boats; boats imply sails; sails imply ropes. And ropes imply hemp. I sometimes wonder how many of our own prized scientific arguments will appear equally foolish from the vantage of three centuries.)

A useful index of our knowledge about a planet is the number of bits of information necessary to characterize what we know of its surface—in effect the number of black and white dots in halftone photographic reproductions summarizing all existing imagery. Back in Huygens' day about 10 bits of information, all obtained by brief glimpses

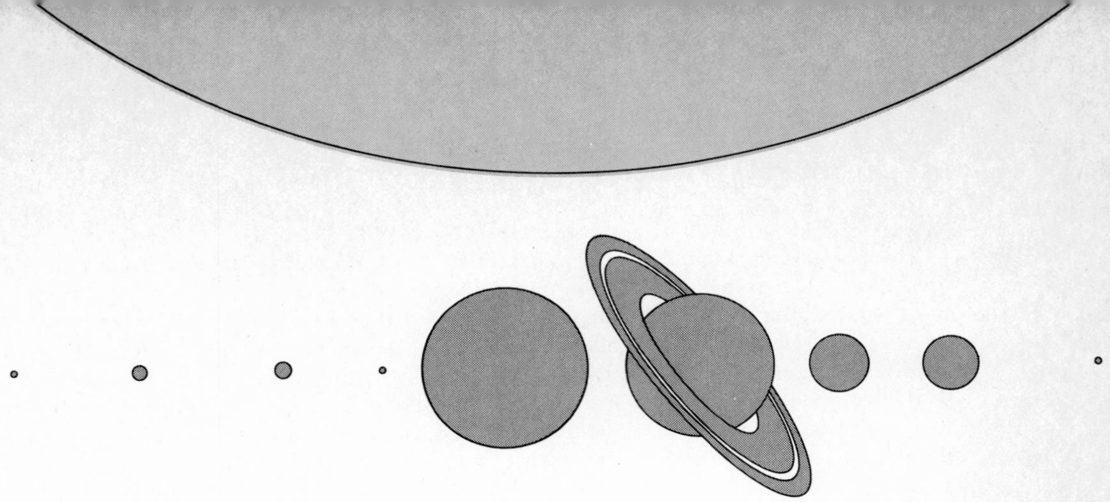

| | MERCURY | VENUS | EARTH | MARS | JUPITER | SATURN | URANUS | NEPTUNE | PLUTO |
|---|---|---|---|---|---|---|---|---|---|
| MAXIMUM DISTANCE FROM SUN (MILLIONS OF KILOMETERS) | 69.7 | 109 | 152.1 | 249.1 | 815.7 | 1,507 | 3,004 | 4,537 | 7,375 |
| MINIMUM DISTANCE FROM SUN (MILLIONS OF KILOMETERS) | 45.9 | 107.4 | 147.1 | 206.7 | 740.9 | 1,347 | 2,735 | 4,456 | 4,425 |
| MEAN DISTANCE FROM SUN (MILLIONS OF KILOMETERS) | 57.9 | 108.2 | 149.6 | 227.9 | 778.3 | 1,427 | 2,869.6 | 4,496.6 | 5,900 |
| MEAN DISTANCE FROM SUN (ASTRONOMICAL UNITS) | .387 | .723 | 1 | 1.524 | 5.203 | 9.539 | 19.18 | 30.06 | 39.44 |
| PERIOD OF REVOLUTION | 88 DAYS | 224.7 DAYS | 365.26 DAYS | 687 DAYS | 11.86 YEARS | 29.46 YEARS | 84.01 YEARS | 164.8 YEARS | 247.7 YEARS |
| ROTATION PERIOD | 59 DAYS | −243 DAYS RETROGRADE | 23 HOURS 56 MINUTES 4 SECONDS | 24 HOURS 37 MINUTES 23 SECONDS | 9 HOURS 50 MINUTES 30 SECONDS | 10 HOURS 14 MINUTES | −11 HOURS RETROGRADE | 16 HOURS | 6 DAYS 9 HOURS |
| ORBITAL VELOCITY (KILOMETERS PER SECOND) | 47.9 | 35 | 29.8 | 24.1 | 13.1 | 9.6 | 6.8 | 5.4 | 4.7 |
| INCLINATION OF AXIS | <28° | 3° | 23°27′ | 23°59′ | 3°05′ | 26°44′ | 82°5′ | 28°48′ | ? |
| INCLINATION OF ORBIT TO ECLIPTIC | 7° | 3.4° | 0° | 1.9° | 1.3° | 2.5° | .8° | 1.8° | 17.2° |
| ECCENTRICITY OF ORBIT | .206 | .007 | .017 | .093 | .048 | .056 | .047 | .009 | .25 |
| EQUATORIAL DIAMETER (KILOMETERS) | 4,880 | 12,104 | 12,756 | 6,787 | 142,800 | 120,000 | 51,800 | 49,500 | 6,000 (?) |
| MASS (EARTH = 1) | .055 | .815 | 1 | .108 | 317.9 | 95.2 | 14.6 | 17.2 | .1 (?) |
| VOLUME (EARTH = 1) | .06 | .88 | 1 | .15 | 1,316 | 755 | 67 | 57 | .1 (?) |
| DENSITY (WATER = 1) | 5.4 | 5.2 | 5.5 | 3.9 | 1.3 | .7 | 1.2 | 1.7 | ? |
| OBLATENESS | 0 | 0 | .003 | .009 | .06 | .1 | .06 | .02 | ? |
| ATMOSPHERE (MAIN COMPONENTS) | NONE | CARBON DIOXIDE | NITROGEN, OXYGEN | CARBON DIOXIDE, ARGON (?) | HYDROGEN, HELIUM | HYDROGEN, HELIUM | HYDROGEN, HELIUM, METHANE | HYDROGEN, HELIUM, METHANE | NONE DETECTED |
| MEAN TEMPERATURE AT VISIBLE SURFACE (DEGREES CELSIUS) S = SOLID, C = CLOUDS | 350(S) DAY −170(S) NIGHT | −33 (C) 480 (S) | 22 (S) | −23 (S) | −150 (C) | −180 (C) | −210 (C) | −220 (C) | −230(?) |
| ATMOSPHERIC PRESSURE AT SURFACE (MILLIBARS) | $10^{-9}$ | 90,000 | 1,000 | 6 | ? | ? | ? | ? | ? |
| SURFACE GRAVITY (EARTH = 1) | .37 | .88 | 1 | .38 | 2.64 | 1.15 | 1.17 | 1.18 | ? |
| MEAN APPARENT DIAMETER OF SUN AS SEEN FROM PLANET | 1°22′40″ | 44′15″ | 31′59″ | 21′ | 6′09″ | 3′22″ | 1′41″ | 1′04″ | 49″ |
| KNOWN SATELLITES | 0 | 0 | 1 | 2 | 13 | 10 | 5 | 2 | 0 |
| SYMBOL | ☿ | ♀ | ⊕ | ♂ | ♃ | ♄ | ⛢ | ♆ | ♇ |

through telescopes, would have characterized man's knowledge of the surface of Mars. By the time of the close approach of Mars to the earth in 1877 that number had risen to perhaps a few thousand (if we exclude a large amount of erroneous information, such as the drawings of "canals" that we now know were entirely illusory). With further visual observations and the rise of astronomical photography the amount of information grew slowly until the advent of space-vehicle exploration of the planet provided a surge of new data. Just 22 photographs obtained in 1965 by the *Mariner 4* flyby mission represented five million bits of information, roughly comparable to all previous photographic knowledge of the planet, although they covered only a tiny fraction of the planet's area. The dual flyby mission of *Mariner 6* and *Mariner 7* in 1969 extended the coverage, increasing the bit total by a factor of 100, and in 1971 and 1972 the *Mariner 9* orbiter increased it by another factor of 100. The *Mariner 9* photographic results from Mars correspond roughly to 10,000 times the total previous photographic knowledge of Mars gathered over the history of mankind. The infrared and ultraviolet spectroscopic data and other information obtained by *Mariner 9* represent a similar enhancement.

The vast amount of new photographic information involves not only an advance in coverage, or quantity, but also a spectacular advance in resolution, or quality. Before the voyage of *Mariner 4* the smallest feature reliably detected on the surface of Mars was several hundred kilometers across. With the completion of the *Mariner 9* mission several percent of the planet's area has been observed at an effective resolution of 100 meters, an improvement in resolution by a factor of 1,000 in the past 10 years and by a factor of 10,000 since Huygens' time. It is only because of this improvement in resolution that we know of vast volcanoes, laminated polar formations, sinuous channels, great rift valleys, dune fields, crater-associated dust streaks and

many other instructive and mysterious features of the Martian environment.

Both resolution and coverage are required in order to provide adequate information about a newly explored planet. For example, by an unlucky coincidence the *Mariner 4, Mariner 6* and *Mariner 7* spacecraft observed the old, cratered and comparatively uninteresting part of Mars and gave no hint of the young and geologically active third of the planet that was revealed by *Mariner 9.* Intelligent life on the earth would be entirely undetectable by photography in reflected sunlight unless about 100-meter resolution was achieved, at which point the urban and agricultural geometry of our technological civilization would become strikingly evident. This means that if there had been a civilization on Mars comparable in extent and level of development to our own, it would not have been detected photographically until the *Mariner 9* mission. There is no reason to expect such civilizations on other planets in our solar system; my point is that we are only now beginning an adequate reconnaissance of our neighboring worlds. There is no question that astonishments and delights await us as both resolution and coverage are dramatically improved in photography, and in spectroscopic and other methods, by future space-probe missions.

The vigor of the burgeoning planetary sciences and the volume and detail of recent findings will impress anyone who attends a meeting of the Division for Planetary Sciences of the American Astronomical Society. At the 1975 meeting in February there were reports on the discovery of water vapor in the atmosphere of Jupiter, of ethane on Saturn, of possible hydrocarbons on the asteroid Vesta, of an atmospheric pressure approaching that of the earth on Saturn's moon Titan and of radio bursts in the decameter-wavelength range from Saturn. Jupiter's moon Ganymede had been detected by radar, and the radio-emission spectrum of another Jovian moon, Callisto, had been elaborated.

And spectacular new views of Jupiter and Mercury and their magnetospheres were presented by the *Pioneer 11* and *Mariner 10* experimenters.

Such discoveries are important and exciting in themselves, but it is their implications and interrelations that are most significant. Every new finding adds to the accumulation of evidence that is required before we can write an authentic history of the origin and evolution of the solar system. No complete version of that history has yet been accepted, but this field of study is now rich in provocative hints and ingenious surmises. Apart from an understanding of the solar system as a whole, it is becoming clear that information about any planet or satellite illuminates our knowledge of the others. In particular, if we are to understand the earth, we must have a comprehensive knowledge of the other planets. Let me give a few examples of what might be called comparative planetology.

There is now observational evidence to support an idea I first proposed in 1960: that the high temperatures on the surface of Venus are due to a runaway "greenhouse effect" in which water and carbon dioxide in the planetary atmosphere impede the emission of thermal radiation from the surface to space. The surface temperature rises to the point where there is an equilibrium between the visible sunlight arriving at the surface and the infrared radiation leaving it; this higher surface temperature results in a higher vapor pressure of the greenhouse gases, carbon dioxide and water, and the process continues until all the carbon dioxide and water is in the vapor phase, producing a planet with a high atmospheric pressure and a high surface temperature. The reason Venus has such an atmosphere and the earth does not seems to be that Venus receives a little more sunlight than the earth. If the sun were to become brighter or the earth's surface and clouds were to become darker, could the earth become a replica of this classical vision of hell? Venus may be a cautionary tale for our technical civilization, which has the capability to profoundly alter the environment of our small planet.

In spite of the expectations of almost all planetary scientists, Mars turns out to be covered with thousands of sinuous, tributaried channels that are probably one or two billion years old. Whether they were formed by running water or by running carbon dioxide, such channels could not be carved under present atmospheric conditions; they require

**MAIN PROPERTIES** of the planets are summarized in table on the opposite page. Drawings at top show the planets' sizes with respect to the sun. Minus sign in front of the rotation period of Venus and Uranus indicates that those planets rotate in a direction opposite to the direction in which the other planets rotate. Eccentricity of a planet's elliptical orbit is customarily expressed as the distance between the two foci divided by the length of the major axis. Oblateness, the amount by which a rotating body is flattened, is given as the difference between the equatorial and the polar diameters divided by the equatorial diameter of body. Data that are presented in this chart and the one on the next page were compiled from a number of sources with the assistance of Jay D. Goguen of Cornell University.

| NAME OF SPACECRAFT | DATE OF LAUNCH | DESTINATION | DATE OF ENCOUNTER | NEAREST APPROACH (KILOMETERS) | STATUS OF MISSION |
|---|---|---|---|---|---|
| VENERA 1 | 2/12/61 | VENUS | — | 100,000 | Radio contact lost 7.5 million kilometers from earth. |
| MARINER 1 | 7/22/62 | VENUS | — | — | Booster rocket deviated from course and was destroyed by range safety officer. |
| MARINER 2 | 8/26/62 | VENUS | 12/14/62 | 35,000 | First flyby of another planet; found high temperature (400 degrees Celsius) arises from surface, not atmosphere; no evidence of magnetic field. |
| MARS 1 | 11/1/62 | MARS | — | 190,000 | Radio contact lost 106 million kilometers from earth. |
| ZOND 1 | 4/2/64 | VENUS (?) | — | — | Radio contact lost within month after launch. |
| MARINER 3 | 11/5/64 | MARS | — | — | Shroud failed to jettison; radio contact lost soon after launch. |
| MARINER 4 | 11/28/64 | MARS | 7/14/65 | 10,000 | First flyby of Mars; returned 22 television pictures of Martian surface, other data. |
| ZOND 2 | 11/30/64 | MARS | — | — | Radio contact lost 5/2/65. |
| VENERA 2 | 11/12/65 | VENUS | 2/27/66 | 24,000 | Passed Venus but failed to return data. |
| VENERA 3 | 11/16/65 | VENUS | 3/1/66 | LANDED | First spacecraft to land on another planet; failed to return data. |
| VENERA 4 | 6/12/67 | VENUS | 10/18/67 | LANDED | First on-site measurements of temperature, pressure and composition of Venusian atmosphere; probe transmitted data during 94-minute parachute descent. |
| MARINER 5 | 6/14/67 | VENUS | 10/19/67 | 4,000 | Measured structure of upper atmosphere of Venus during flyby. |
| VENERA 5 | 1/5/69 | VENUS | 5/16/69 | LANDED | Probes transmitted data on pressure, temperature and composition of atmosphere during parachute descent; missions similar to that of *Venera 4*. First successful landing on another planet. |
| VENERA 6 | 1/10/69 | VENUS | 5/17/69 | LANDED | |
| MARINER 6 | 2/25/69 | MARS | 7/31/69 | 3,390 | Flyby obtained infrared and ultraviolet spectra of atmosphere; returned 76 pictures of surface, other data. |
| MARINER 7 | 3/27/69 | MARS | 8/5/69 | 3,500 | Mission identical with that of *Mariner 6*; returned 126 pictures of surface, 33 of south-polar region. |
| VENERA 7 | 8/17/70 | VENUS | 12/15/70 | LANDED | Mission similar to those of *Venera 4*, *Venera 5* and *Venera 6*. |
| MARINER 8 | 5/8/71 | MARS | — | — | Malfunctioned during launch; crashed in Atlantic. |
| MARS 2 | 5/19/71 | MARS | 11/27/71 | LANDED | Orbiter achieved Mars orbit; lander crashed to surface. |
| MARS 3 | 5/28/71 | MARS | 12/2/71 | LANDED | Orbiter achieved Mars orbit and returned data; descent module soft-landed and transmitted 20 seconds of featureless television data before failing. |
| MARINER 9 | 5/30/71 | MARS | 11/13/71 | 1,395 | First spacecraft to go into orbit around another planet; returned 7,329 pictures of surface, atmosphere, clouds and satellites, together with other data. |
| PIONEER 10 | 3/3/72 | JUPITER | 12/4/73 | 131,400 | Successfully traversed asteroid belt; investigated interplanetary medium, Jovian magnetosphere and atmosphere; returned more than 300 pictures of Jovian clouds and satellites; first spacecraft to use gravity-assisted trajectory; first man-made object to escape solar system. |
| VENERA 8 | 3/26/72 | VENUS | 7/22/72 | LANDED | Survived Venusian surface conditions for 50 minutes; determined radioactive content of surface; on entry measured winds and sunlight penetrating clouds. |
| PIONEER 11 | 4/6/73 | JUPITER SATURN | 12/3/74 (J) 9/79 (S) | 46,400 (J) | Second Jupiter flyby; now en route to Saturn, then to leave solar system. |
| MARS 4 | 7/21/73 | MARS | 1/74 | ? | Went into orbit around Mars; returned photographs of surface and other data. |
| MARS 5 | 7/25/73 | MARS | 1/74 | ? | |
| MARS 6 | 8/5/73 | MARS | 2/74 | LANDED | Descent module failed at touchdown; entry data suggest high argon content of atmosphere. |
| MARS 7 | 8/9/73 | MARS | — | — | Radio contact lost 3/12/74. |
| MARINER 10 | 11/3/73 | VENUS MERCURY | 2/5/74 (V) 3/29/74 (M) | 5,800 (V) 700 (M) | First probe of Mercury; returned more than 8,000 pictures and other data from Venus and Mercury; re-encountered Mercury 9/21/74 and 3/16/75. |
| VIKING 1 | 8/75 | MARS | 6/76 | TO LAND | Orbiter to study atmosphere and photograph surface; lander to study atmosphere at surface, investigate surface geology and chemistry and test soil for signs of extraterrestrial life. |
| VIKING 2 | 9/75 | MARS | 8/76 | TO LAND | |
| MARINER 11 | 8/77 | JUPITER SATURN | 1979 (J) 1981 (S) | ? | To conduct comparative studies of two outer planets and their 23 satellites; to investigate nature of Saturn's rings; to measure interplanetary medium out to Saturn's orbit; 20,000 photographs planned. |
| MARINER 12 | 9/77 | JUPITER SATURN | 1979 (J) 1981 (S) | ? | |
| PIONEER 12 | 5/78 | VENUS | 12/78 | TO LAND | Orbiter to study interaction of atmosphere with solar wind over one 243-day period; "bus" to drop three small probes toward surface, then relay data to earth as they enter atmosphere. |
| PIONEER 13 | 8/78 | VENUS | 12/78 | TO LAND | |

much higher pressures and probably higher polar temperatures. And so the channels bear witness to at least one epoch and perhaps many previous epochs of milder conditions on Mars, implying that there have been major climatic variations over the history of the planet. We do not know whether such variations are the result of internal causes or of external ones. If the causes are native to Mars, it becomes important to learn whether the earth might, perhaps even as a result of the activities of man, be subject to climatic excursions of Martian magnitude. If the Martian climatic variations were the result of external causes (perhaps variations in the luminosity of the sun), then a correlation of Martian paleoclimatology and terrestrial paleoclimatology would be extremely interesting.

*Mariner* 9 arrived at Mars in the midst of a great global dust storm, and its data make it possible to determine whether such storms heat a planetary surface or cool it. Any theory with pretensions to predicting the climatic consequences of an increase in the abundance of finely divided particles in the earth's atmosphere had better be able to provide the correct answer for that dust storm on Mars. In fact, drawing on our *Mariner* 9 experience, James B. Pollack of the Ames Research Center of the National Aeronautics and Space Administration and Owen B. Toon and I at Cornell University have calculated the effects of single and multiple volcanic explosions on the earth's climate and have been able to reproduce, within the limits of experimental error, the climatic effects that were observed after actual volcanic explosions. The perspective of planetary astronomy, which alone enables us to view a planet as a whole, seems to be good training for studies of the earth. As another example of the contribution made by planetary studies to terrestrial problems, one of the main groups investigating the effect on the earth's ozone layer of the injection into the atmosphere of fluorocarbon propellants from aerosol cans is one headed by Michael B. McElroy of Harvard University—a group that cut its teeth on the

INTERPLANETARY SPACECRAFT already launched by the U.S. and the U.S.S.R. are listed on the opposite page, along with several projected U.S. missions that are in the advanced planning stage. Excluded are earth-orbiting vehicles, lunar missions and the probes of the interplanetary medium.

physics and chemistry of the atmosphere of Venus.

We now know from space-vehicle observations something of the density of impact craters of different sizes on Mercury, the moon, Mars and its satellites, and radar studies are beginning to provide such information for Venus. Although the surface of the earth has been heavily altered by wind and water and by crustal folding and faulting, we also have some information about craters on the surface of the earth. If the population of objects that produced such impacts were the same for all these planets, it might be possible to work out both the absolute and the relative chronology of various cratered surfaces. The trouble is that we do not yet know whether the impacting objects are from a common source (for example the asteroid belt) or are of local origin (for example rings of debris swept up in the final stages of planetary accretion).

The heavily cratered lunar highlands speak to us of an early epoch in the history of the solar system, when the frequency of cratering was much higher than it is today; the present population of interplanetary debris fails by a large factor to account for the density of the highland craters. On the other hand, the lunar maria, or "seas," show a much lower crater density, which can be explained quite well by the present population of interplanetary debris: mostly asteroids and possibly dead comets. For planetary surfaces that are not so heavily cratered it is possible to determine something of the absolute age, a great deal about the relative age and in certain cases even something about the distribution of sizes in the population of objects that made the craters. On Mars, for example, we find that the flanks of the large volcanic mountains are almost free of impact craters, implying their comparative youth: they have not been around long enough to have accumulated much in the way of impact scars. That is the basis for the hypothesis of comparatively recent Martian volcanism.

The ultimate objective of comparative planetology, it might be said, is something like a vast computer program into which we insert a few input parameters (perhaps the initial mass, composition and angular momentum of a protoplanet and the population of neighboring objects that strike it) and then derive the complete evolution of the planet. We are far from having such a deep understanding of planetary evolution at present, but we are much closer than would have

been thought possible only a few decades ago.

In addition every new set of discoveries raises a host of questions we were not until now even able to ask. I shall mention just a few of them.

The initial radar glimpse of the craters of Venus shows them as being extremely shallow. There is no liquid water to erode Venus' surface, and the lower atmosphere seems to move so slowly that its winds may not be strong enough to fill the craters with dust. Could the craters of Venus be filled by the slow collapse of very slightly molten walls, flowing like pitch?

The most popular explanation for the generation of planetary magnetic fields invokes rotation-driven convection currents in an electrically conducting planetary core. Mercury, which rotates only once every 59 days, was expected to have no detectable magnetic field, but *Mariner 10* discovered one. Apparently a serious reappraisal of theories of planetary magnetism is in order.

Only Saturn has rings. Why?

There is an exquisite array of longitudinal sand dunes on Mars, nestling against the interior ramparts of the large eroded crater Procter. In Colorado, in the Great Sand Dunes National Monument, similar sand dunes nestle in a curve of the Sangre de Cristo Mountains. The Martian dunes and the terrestrial ones have the same total extent, the same dune-to-dune spacing and the same dune heights. Yet the Martian atmospheric pressure is only a two-hundredth of the pressure on the earth, so that the winds needed to push the sand grains around must be 10 times stronger than those on the earth; moreover, the distribution of particle sizes may be quite different on the two planets. How then can the dune fields produced by windblown sand be so similar?

Observations made from *Mariner 9* imply that the winds on Mars at least occasionally exceed half the local speed of sound. Are the winds ever much stronger? And if they are, what is the nature of a transonic meteorology?

There are pyramids on Mars that are about three kilometers across at the base and one kilometer high. They are not likely to have been constructed by Martian pharaohs. The rate of sandblasting by wind-transported grains on Mars is perhaps 10,000 times greater than the rate on the earth because of the greater speeds necessary to move particles in the thinner Martian atmosphere. Could the facets of the Martian pyramids have

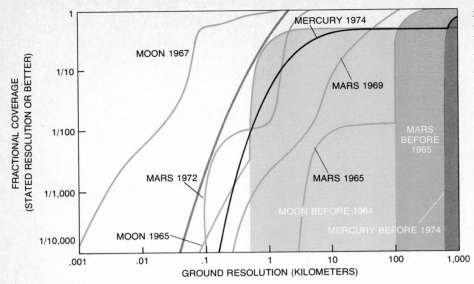

**INCREASE IN KNOWLEDGE** about the surfaces of the moon, Mars and Mercury resulting from the space missions of the past few years is estimated on this graph in terms of both coverage (*vertical scale*) and resolution (*horizontal scale*). Mercury is represented by only one curve, since prior to the *Mariner 10* mission of 1974 no object smaller than about 800 kilometers could be resolved in photographs of its surface. For the purpose of comparison, gray band indicates resolution where any work of man would be detectable on the earth.

been eroded by millions of years of such sandblasting from more than one prevailing wind direction?

The moons of the outer solar system are almost certainly not replicas of our own rather dull satellite. Many of them have such a low density that they must consist largely of ices of methane, ammonia or water. What will their surfaces be like close up? How do impact craters erode on an icy surface? Might there be volcanoes of solid ammonia with lavas of liquid ammonia trickling down their sides? Why is Io, the innermost large satellite of Jupiter, enveloped in a cloud of gaseous sodium? Why is one side of Saturn's moon Iapetus six times brighter than the other? Is it because of a particle-size difference? A chemical difference? How did such differences become established, and why did they become established on Iapetus and nowhere else in the solar system in such a symmetrical way? The gravity of Saturn's largest moon, Titan, is low enough and the temperature of the upper atmosphere is high enough for the hydrogen in the atmosphere to escape rapidly into space. Yet the spectroscopic evidence suggests that a substantial quantity of hydrogen remains on Titan. Why?

Beyond Saturn the solar system is still almost literally clouded in ignorance. Our feeble telescopes have not even reliably determined the periods of rotation of Uranus, Neptune and Pluto, much less the character of their clouds and at-mospheres and the nature of their satellite systems.

One of the most tantalizing issues, and one that we are just beginning to approach seriously, is the question of organic chemistry and biology elsewhere in the solar system. The issue of whether there are organisms both large and small, on Mars in particular, is entirely open. The Martian environment is by no means so hostile as to exclude life, but we do not know enough about the origin and evolution of life to guarantee its presence there—or anywhere else. The three microbiology experiments, the organic chemistry experiment and the camera systems aboard the two Viking vehicles scheduled to land on Mars next summer may provide the first experimental evidence on the matter. The hydrogen-rich atmospheres of places such as Jupiter, Saturn, Uranus and Titan are in significant respects similar to the atmosphere of the earth at the time of the origin of life. From laboratory simulation experiments we know that organic molecules are synthesized in high yield under those conditions. (In the atmospheres of Jupiter and Saturn such molecules would be carried by convection to depths where they would be decomposed by heat, but even there the steady-state concentration of organic molecules may be significant.) In all simulation experiments the application of energy to such atmospheres produces a brownish polymeric material that in many respects resembles the brownish coloring matter in the clouds of Jupiter and Saturn. Titan may be completely covered with a brownish organic material. It is possible that the next few years will see major and unexpected discoveries in the infant science of exobiology.

The principal means for the continued exploration of the solar system over the next decade or two will surely be unmanned planetary missions. Scientific space vehicles have now been launched successfully to all the planets known to the ancients. If even a small fraction of the missions that are scheduled and have been proposed are implemented, it is clear that the present golden age of planetary exploration will continue.

Yet even a preliminary reconnaissance of the entire solar system out to Pluto and the more detailed exploration of a few planets (by, for example, vehicles that will traverse the surface of Mars or penetrate the atmosphere of Jupiter) will not solve the fundamental problem of solar-system origins. What we need is to discover other solar systems, perhaps at various stages in their evolution. Advances in ground-based and space-borne instruments over the next two decades may make it possible to detect dozens of planetary systems around nearby single stars. Recent observational studies of multiple-star systems by Helmut Abt and Saul Levy of the Kitt Peak National Observatory suggest that as many as a third of all stars have planetary companions. We do not know whether such systems will be like ours or will be built on very different principles. Richard Isaacman of Cornell and I have calculated a range of possible planetary systems based on a theoretical model originally devised by Stephen H. Dole of the Rand Corporation. The assumptions behind these models are so simple as to make us believe they are unrealistic, and yet the range of systems to which they give rise is intriguing. The time may not be far off when we shall have observational information on the distribution in space of various types of planetary systems. We may then be able to echo Huygens: "What a wonderful and amazing Scheme we have here of the magnificent Vastness of the Universe! So many Suns, so many Earths!"

Centuries hence, when current social and political problems may seem as remote as the problems of the Thirty Years' War are to us, our age may be remembered chiefly for one fact: It was the time when the inhabitants of the earth first made contact with the vast cosmos in which their small planet is embedded.

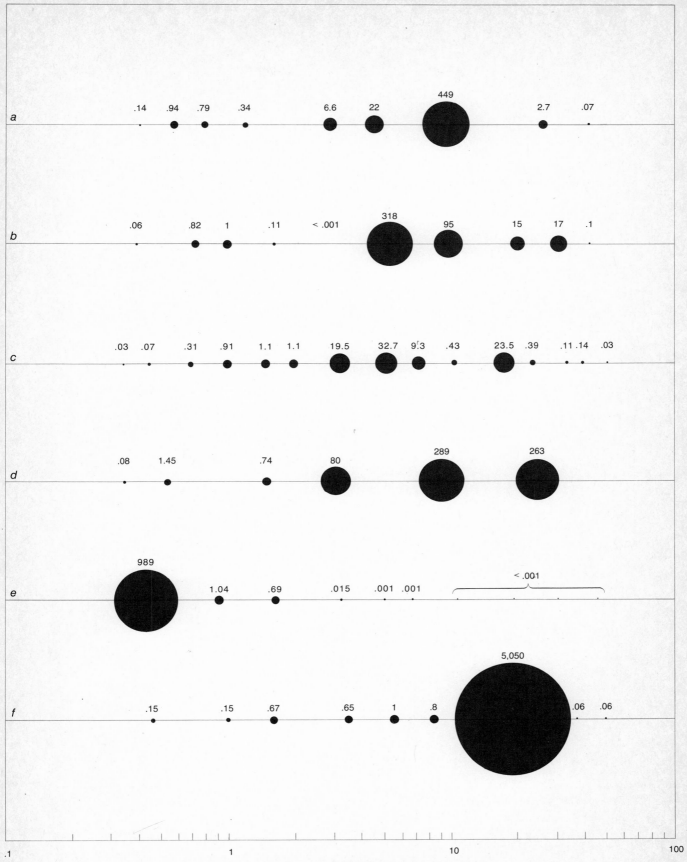

ALTERNATIVE PLANETARY SYSTEMS were calculated by the author and his colleague Richard Isaacman of Cornell on the basis of a theoretical model originally devised by Stephen H. Dole of the Rand Corporation. The assumptions behind such calculations are probably too simple at present, but the exercise is thought to be suggestive. Numbers above hypothetical planets denote mass in multiples of earth's mass. Horizontal scale indicates semimajor axes of planets' elliptical orbits. System labeled *b* is our solar system.

# 2

# THE ORIGIN AND EVOLUTION OF THE SOLAR SYSTEM

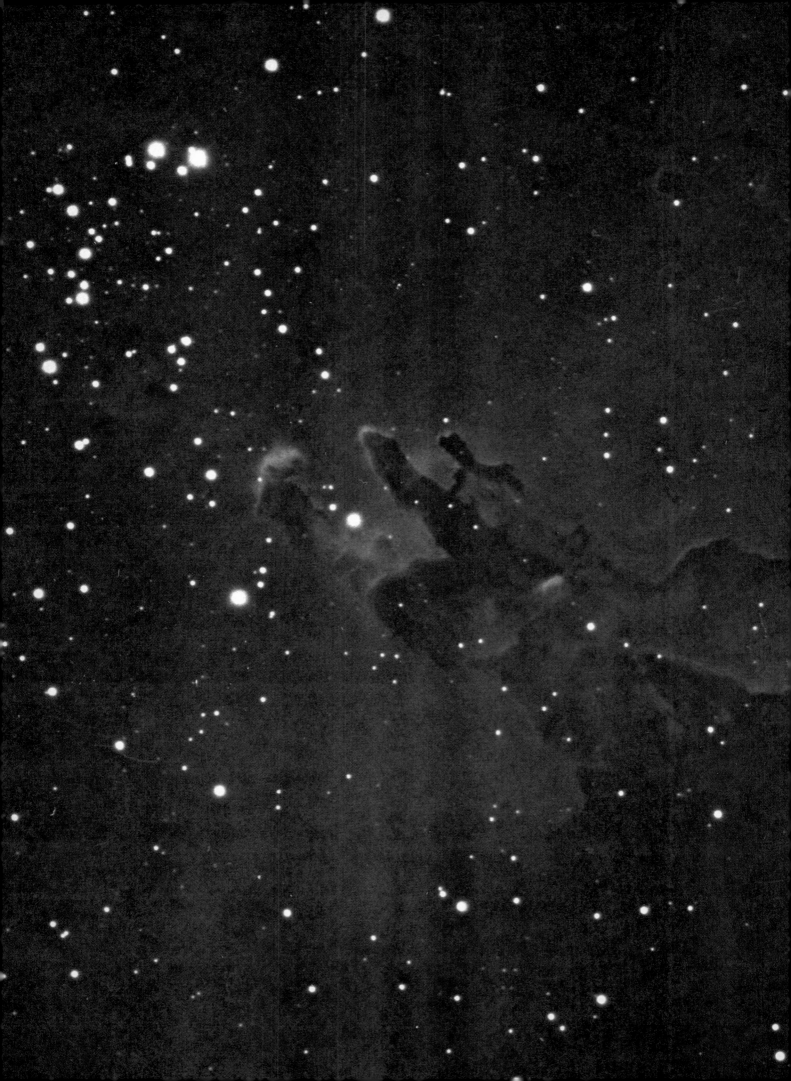

# The Origin and Evolution of the Solar System

A. G. W. CAMERON

*It is generally agreed that some 4.6 billion years ago
the sun and the planets formed out of a rotating
disk of gas and dust. Exactly how they did
so remains a lively topic of investigation*

A great cloud of gas and dust contracted through interstellar space 4.6 billion years ago, far out along one of the curved arms of our spiral galaxy. The cloud collapsed and spun more rapidly, forming a disk. At some stage a body collected at the center of the disk that was so massive, dense and hot that its nuclear fuel ignited and it became a star: the sun. At some stage the surrounding dust particles accreted to form planets bound in orbit around the sun and satellites bound in orbit around some of the planets.

So goes—in very broad outline—the nebular hypothesis of the origin of the solar system. Its central idea was proposed more than 300 years ago. It sounds simple enough, and it makes intuitive sense to the layman; indeed, some version of it is accepted by most astronomers today. And yet beyond the broad outlines there is no consensus among students of the origin and evolution of the solar system. We still have no generally accepted theory to explain how the primitive solar nebula formed, how and when the sun began to shine and how and when the planets coalesced out of swirling dust.

It was René Descartes who first proposed (in 1644) the concept of a primitive solar nebula: a rotating disk of gas and dust out of which the planets and their satellites are made. A century later (in 1745) Georges Louis Leclerc de Buffon put forward a second theory: that a massive body (he suggested a comet) came close to the sun and ripped out of it the material that constituted the planets and their satellites. In the two centuries after Buffon the many theories that were propounded tended to follow in the tradition of either Descartes's monistic view or Buffon's dualistic one; the balance of favor swung back and forth between them. The most significant early monistic theories were those of Immanuel Kant and Pierre Simon de Laplace, who elaborated on Descartes's original idea by explaining how the cloud of gas and dust, shrinking to form the sun, would have spun faster and faster because of the conservation of angular momentum: a decrease in the radius of a rotating mass must be balanced by an increase in its rotational speed. Laplace suggested that a series of rings were shed, from whose dust the planets and satellites were formed. At the end of the 19th century dissatisfaction with the ability of the nebular hypothesis to explain the accretion of matter into the planets brought dualistic theories back into favor. Today they have been generally abandoned; it

seems clear that most of the material that might have been drawn out of the sun by, say, the approach of another star would have fallen back into the sun or dispersed in space before any solid condensates could coagulate into planets.

A major reason for the wide range of early theories of the origin of the solar system was the lack of observational data—of facts to be explained by a theory. The history of the earth's first few hundreds of millions of years is missing from the geological record, which could therefore offer no clues to the environment in which this sample of a planet was born, and the limited capabilities of telescopes restricted the astronomical data. The early theories were devised to explain only a few observations: the spacing of the planetary orbits increased in a regular way (in accordance with what is known as Bode's law); planetary orbital motions and spins tended to have the same direction of rotation; the sun accounted for only a small fraction of the total angular momentum of the solar system, even though it accounted for the greatest fraction by far of the total mass of the system. These few facts provided few constraints on theory, and so the theories proliferated.

In just the past three decades the situation has changed dramatically. We have a vast amount of new information that imposes additional and powerful constraints on any theory. The new knowledge stems notably from new research on meteorites and from the data returned to the earth by spacecraft dispatched to other bodies in the solar system.

The meteorites are samples of primitive solar-system material. They are evi-

NEW STARS ARE BORN, as the sun may have been born, in gaseous emission nebulas: diffuse, dusty clouds of hot interstellar gas. The photograph on the opposite page, made with the Mayall four-meter reflecting telescope at the Kitt Peak National Observatory in Arizona, shows the nebula designated M16 or NGC 6611, called the Eagle Nebula, in the local arm of our galaxy. Within the nebula clouds of gas have condensed relatively recently to form bright blue-white stars, and other such clouds are still condensing. Ultraviolet radiation from the hot new stars ionizes hydrogen atoms in the remaining gas, giving rise to free electrons and protons. When high-energy electrons recombine with protons, light is emitted at the hydrogen-alpha wavelength of the spectrum: red light that illuminates the cloud and silhouettes dense, dusty, cooler regions of the nebula where the light does not penetrate.

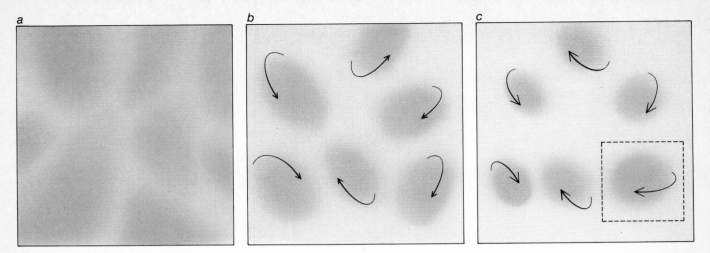

GALAXIES FORMED in the thin, expanding primordial gas (mostly hydrogen, with some helium) when regions of somewhat greater density (*a*) contracted gravitationally to form protogal- axies (*b*), rotating because of the net effect of gas eddies within them. The protogalaxies continued to contract gravitationally, and then to rotate faster (*c*). One of them (*rectangle*) was our own.

dently fragments of rather small bodies that have collided and broken up, sending many of their pieces into new orbits that ultimately intersect the earth. They bring to us, trapped in their interior, samples of the gases of the solar nebula. The details of their mineralogy provide clues to the temperatures and pressures in the nebula at the time its individual grains were last exposed to chemical re-action with its gases. From the relative amounts of the products of radioactive decay that remain trapped in the interior of the meteorites we learn how long ago the original elements that gave rise to certain radioactive isotopes were as-sembled to form the meteorites' parent bodies.

One of the primary scientific goals of the space-probe program was to advance understanding of the origin of the solar system, and the program has already borne fruit. Measurements made by spacecraft have refined our knowledge of planetary masses and radii, from which we derive accurate mean densities of the planets and clues to their internal composition. By observing how the grav-itational potentials of a planet differ from those of a perfectly uniform sphere we derive constraints on the degree to which the density can vary in different parts of the planet's interior. Determin-ing whether or not a planet has an in-trinsic magnetic field tells us something about the planet's internal dynamics. Spacecraft data on the composition of a planetary atmosphere reveal something about the gases that once were incor-porated in the planet and about chemi-cal interactions between the atmosphere and the planet's surface. Examining the incredibly detailed images of solid plane-

tary surfaces that have been sent back by spacecraft cameras, we can see how volcanic and other geological processes have operated on other planets. The density of craters tells us about the ter-minal stages of the planet's accretion and about the numbers of smaller bodies that have wandered through the solar system.

Still other constraints come from the general advances in astrophysics that have marked the past three decades. We now know that our galaxy as a whole is between two and three times older than the solar system; we therefore have good reason to believe that the conditions we see today in the galaxy are not very dif-ferent from those at the time the solar system was formed. We see regions in our galaxy in which stars have been formed in the recent past and are prob-ably still being formed today; that gives us important information if we believe the sun and the solar nebula formed as parts of the same general process. We have learned much about the birth and death of stars and how elements origi-nate in nuclear reactions within explod-ing stars and are formed into tiny grains of interstellar dust, and about how those grains concentrate in the dark patches in the sky that blot out the light coming to us from distant stars. Those grains of dust and the interstellar gases that ac-company them were the raw material of the solar nebula. Let me now try to weave the many threads of information into a coherent picture of the solar sys-tem's formation.

Galaxies form when gas—mostly hy-drogen—collapses out of intergalac-tic space. Many billions of years before

the origin of the solar system our galaxy began to take shape in that way. Out of the collapsing gas a first generation of stars was born—stars that still remain spherically distributed around the cen-ter of the galaxy, a reminder of its origi-nal roughly spherical shape. After those first stars were formed the residual gas, because of its intrinsic angular momen-tum, settled into the thin disk that is a characteristic feature of all spiral gal-axies, and further generations of stars formed from the gas in the disk. The more massive of them evolved quickly, forming heavy elements that were eject-ed into the interstellar gas. Some of the heavy elements condensed into tiny grains: the interstellar dust. When enough stars had formed in the central plane of the galaxy, an instability de-

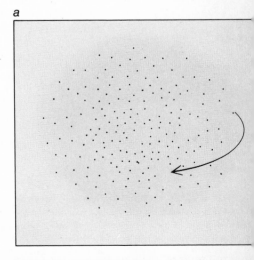

OUR GALAXY EVOLVED as dense lumps of gas contracted within the protogalaxy to form a first generation of stars (*a*). In time the residual gas settled into a disk in the

veloped in their motions that allowed them to cluster together temporarily, forming the spiral arms.

Such arms represent local enhancements of star-population density in the disk; the arms are continuing features that rotate around the center of the galaxy, but the material that constitutes them keeps changing: individual stars spend only about half of their time in one arm before moving on to the next one. Like the stars, the interstellar gas and dust spend about as much time in an arm as they do flowing through the larger spaces between successive arms; the result is that the density of gas and dust is considerably enhanced in a spiral arm. We know from studying galaxies other than our own that it is in these high-density spiral arms that the new stars of spiral galaxies are formed.

Pressure differences arise within the gas, perhaps as the result of a supernova explosion; the gas flows away from regions of higher pressure, but in moving it may tend to pile up somewhere else. Clouds of high-density un-ionized gas accumulate, which typically have a mass of from several hundred to several thousand times the mass of the sun. Gravitational forces tend to pull such a cloud into a more compact configuration. Contraction is opposed, however, by the internal pressure of the gas in the cloud, which tends to make the cloud expand; ordinarily the internal pressure is much stronger than the gravitation and the cloud is in no danger of collapsing. Sometimes, however, a sudden fluctuation in pressure—from a nearby violent event such as a supernova explosion, the formation of a massive star or a large re-

arrangement of the interstellar magnetic field—may compress a cloud to a density much higher than normal. Under such conditions, which are quite rare, gravity may win out over internal pressure, so that the cloud begins to collapse to form stars. As the cloud collapses, its interstellar grains shield its interior against the heating effect of radiation from the stars outside. The temperature of the cloud falls, and the internal pressure becomes less effective. The collapsing cloud breaks into fragments and the fragments break into smaller fragments. When one small fragment eventually completes its collapse, it will have formed into a flattened disk, cool at the edges and very hot at the center: a primitive solar nebula.

What was the nature of the solar nebula and how did it evolve? When did the sun form? Why are there planets? How did they take shape? Quite different pictures of the structure of the primitive solar nebula and of its evolution result from different estimates of its size. Such estimates have usually been arrived at by reasoning backward in time from the present masses of the planets. Let me reproduce such an argument.

In a very general way one can divide the materials of the planets into three classes depending on their volatility: rocky, icy and gaseous. The major constituents of rocky materials are iron and oxides and silicates of magnesium and other metals, notably aluminum and calcium. All these materials would be in solid form at pressures characteristic of the primitive solar nebula and at tem-

peratures in the range from 1,000 to 1,800 degrees Kelvin (degrees Celsius above absolute zero). The four inner planets and the earth's moon (and at least two of the major satellites of Jupiter) appear to be basically rocky. The rocky solids represent about .44 percent by mass of the material out of which the sun formed. The present mass of a rocky planet, then, represents about .44 percent of its share of the primitive solar nebula; the remainder of that share is "missing" because it was too volatile to have been incorporated in the planet.

At a temperature below 160 degrees K. the water in the nebula would be in the form of ice. Ammonia and methane form solids only at a somewhat lower temperature. The ices constitute 1.4 percent by mass of the material out of which the sun formed. Rock-ice mixtures account for most of the mass of Uranus and Neptune and some of the mass of Saturn and Jupiter (and for the bulk of the mass of most of the satellites of the outer planets and the comets). Arguing as for the inner planets, one can assume that the rock and ice now present in such bodies represent 1.4 plus .44 percent, or 1.84 percent, of those bodies' original share of the nebula.

At any temperature likely to have been attained in the primitive solar nebula the very volatile elements—hydrogen and the noble gases such as helium and neon—would remain in the gaseous state. Such gases are incorporated in bodies within the solar system only to the extent that they have been held in planetary atmospheres by gravity and, in the case of hydrogen, held in chemical compounds such as water. (A tiny amount

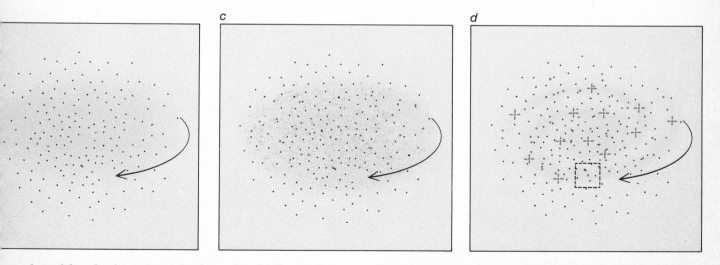

c      d

plane of the galaxy's rotation under the combined influence of gravity and centrifugal force (*b*). Further generations of stars (*color*) formed within the disk (*c*); some stars, evolving rapidly, produced heavier elements through nuclear fusion and ejected them into the disk, where some elements condensed into solid interstellar grains. Instabilities in the motions of gas and stars led to density enhancements that we see as the spiral arms of the galaxy (*d*). The area in the rectangle is enlarged in the first drawing on the next page.

of helium comes from the decay of radioactive elements.) Morris Podolak and I recently analyzed the structure of the outer planets. We determined that hydrogen and helium constitute about 15 percent of Uranus' mass, about 25 percent of Neptune's, about two-thirds of Saturn's and about four-fifths of Jupiter's. In these planets it is necessary to allow for the gaseous components in order to establish the present rock-ice mass.

By thus establishing the rock and the rock-ice masses of the planets and augmenting those masses for the missing constituents that were too volatile to condense it is possible to estimate a minimum mass for the primitive solar nebula: a mass sufficient to account for the formation of the planets. That minimum mass is about 3 percent of the mass of the sun [see *illustration on opposite page*]. (Older estimates arrived at a much smaller mass—less than 1 percent

of the sun's—because they did not allow for enough rock and ice in Jupiter and Saturn.)

The 3 percent figure is definitely a minimum. It assumes that the planets were completely efficient in collecting from the solar nebula all the material that was in condensed form in each planet's orbit in the nebula. For two kinds of solid, however, that collection process might have been quite inefficient. Consider first the tiny unconsolidated grains of interstellar dust, perhaps a micrometer (a thousandth of a millimeter) in diameter, that were not vaporized as the gas-cloud fragment collapsed. The thickness of the nebular disk must have been at least one astronomical unit (the mean distance between the earth and the sun). That dimension is very large compared with the dimensions of any of the planets, which consolidated approximately in the central plane of the disk. Gas-drag effects

would prevent large quantities of these small grains from settling through the nebular gas toward the central plane at a significant rate; if much of the gas was instead dissipated inward to form the sun, the grains would have accompanied the gas and could never have become incorporated in the planets.

Larger bodies (centimeters or meters in diameter), on the other hand, would fall rapidly through the gas toward the midplane but might nevertheless not end up in planets. As a result of a difference between the centrifugal forces that act on the solid bodies and those that act on the gas, the solids would rotate around the central spin axis of the nebula more rapidly than the accompanying gas. They would therefore move through the gas with a relative velocity as high as several hundred miles an hour; a head wind of that speed would tend to slow them down so that they would spiral rather quickly through the gas toward

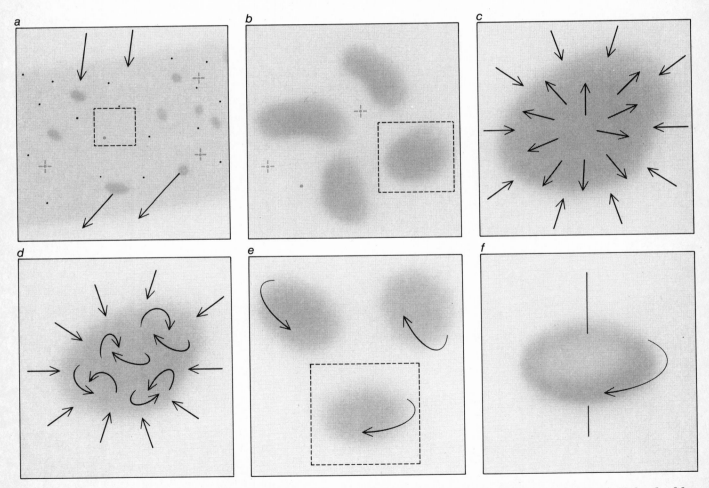

**SOLAR SYSTEM EVOLVED** in a spiral arm about two-thirds of the way out from the center of the galaxy. Stars, gas and dust grains move through the arm, and new stars are born there; massive, short-lived stars outline the arms (*a*). A supernova explosion or the birth of massive stars creates instabilities that concentrate high-density clouds of gas (*b*). Gravitational forces contract the cloud, but the cloud's internal pressure opposes contraction (*c*). If the cloud has enough mass, gravity dominates (*d*) and the cloud collapses. The collapse generates strong gas eddies (*curved arrows*) and breaks the cloud into fragments (*e*); each fragment has a net rotation derived from its major eddies. One of these fragments spins faster and its gas settles into a disk that was the primitive solar nebula (*f*).

the central spin axis and thus be lost to the region of planet formation. For these two reasons the mass of the primitive solar nebula may have been considerably larger than 3 percent of the sun's mass.

Many of the solar-system theories constructed over the past three decades have involved some version of a minimum-mass solar nebula. The concept has a major flaw, however. It assumes that the sun itself was formed directly during the process of collapse and that the primitive solar nebula was marshaled independently around the sun. The trouble is that simple estimates of the amount of angular momentum that must have been contained in the collapsing cloud fragment indicate that it would have been impossible for almost all of the fragment simply to collapse directly to form the sun, leaving a small fringe of nebula to constitute the planets. Such estimates require instead that the nebula's mass be spread out over several tens of astronomical units. The solar nebula itself must have contained substantially more than one solar mass—and probably about two solar masses—of material, with no sun originally present at the central spin axis. Let me first explain the source of the large amount of angular momentum and then show why it indicates that there was not a minimum solar nebula but a massive one.

The strong fluctuations in pressure that led to the rapid compression of the original interstellar cloud and thus brought it to the threshold of gravitational collapse must have stirred the cloud's gases into violent turbulence. Large-scale shearing motions developed —eddies superimposed on eddies, in a wide range of sizes and in many planes and directions. When any one fragment became isolated from such a turbulent cloud, it had a net tendency to spin, derived from the motions of the largest eddies it happened to contain. A fragment's mass, its rate of rotation and its radius combine to endow it with a certain amount of angular momentum, and that momentum must be conserved; as the fragment contracted, it spun faster. The sun turns very slowly, however; in spite of its great mass it accounts for only 2 percent of the solar system's angular momentum. Most of the original angular momentum of the vast quantities of gas that moved in to form the sun must have been transported outward; a considerable part of the original nebula must therefore have remained at

| PLANET | PRESENT MASS (PERCENT OF SUN'S) | AUGMENTED MASS (PERCENT OF SUN'S) |
|---|---|---|
| MERCURY | .000017 | .004 |
| VENUS | .000245 | .056 |
| EARTH | .000304 | .07 |
| MARS | .000032 | .007 |
| JUPITER | .09547 | 1.5 |
| SATURN | .02859 | .77 |
| URANUS | .00436 | .27 |
| NEPTUNE | .00524 | .27 |
| PLUTO | .00025 (?) | .06 (?) |
| TOTAL (MINIMUM MASS OF SOLAR NEBULA) | | 3.0 |

**MINIMUM MASS OF SOLAR NEBULA is estimated by adding up the amount of solar material that must have been present (*column at right*) to account for the present mass (*middle column*) of each planet. Solar-nebula mass thus estimated is 3 percent of mass of sun.**

great distances from the sun to take up that angular momentum.

An additional reason for postulating a massive solar nebula is the observation that young stars tend to lose mass at a prodigious rate early in their lifetime; the loss comes as they pass through what is called their T Tauri stage, which I shall discuss in a bit more detail below. The combination of the mass that remained in the solar nebula and never became part of the sun and the mass that was once in the sun but was lost in the early sun's T Tauri stage could easily have amounted to as much as one solar mass.

As a result of this kind of reasoning—in effect arguing forward from what is known of the principles of star formation rather than backward from the masses of the present planets—Milton R. Pine and I constructed some numerical models of the massive solar nebula. The models extended out to a radial distance of about 100 astronomical units and contained two solar masses of material. In a typical model the temperature was about 3,000 degrees K. near the spin axis and decreased to a few hundred degrees in the region of planet formation. Such temperatures are considerably higher than the temperatures that characterized the collapse of the original interstellar cloud; they develop in the later stages of compression of the gas, once its density becomes high enough so that its own cooling radiation can no longer escape easily. The escape of this radiation is impeded, however, only during the rapid final stages of the collapse; once the gas stops contracting— once the primitive solar nebula is formed

—the radiation can escape relatively quickly, so that in the region of planet formation the nebula will lose most of its heat energy in only a few hundred or a few thousand years.

Such a short cooling time (short compared with the time required to form a sun and planets) presents a difficulty for the massive-nebula model. As the nebula cools it will flatten into a thinner disk, and thin disks have been shown to be dynamically unstable: they tend to deform into a barlike configuration. (Such a deformation might well be the mechanism by which close pairs of double stars are formed, but that evidently did not happen in the solar system.)

There is another time-scale problem for the massive-nebula model. An important process for transporting angular momentum away from the central spin axis so that gas can shrink toward that axis is probably a system of fast meridional currents: gas currents that flow in a plane parallel to the spin axis and at a right angle to the central plane of the nebula. Pine and I estimated that the characteristic time for the outward transport of the angular momentum shed by the inner parts of the primitive solar nebula would be only a few thousand years. John Stewart of the Max Planck Institute for Physics and Astrophysics in Munich has shown that gas turbulence must play an important role in a primitive solar nebula and may cause an even more rapid outward transport of angular momentum.

Both the time for cooling and the time for angular-momentum transport seem too short compared with the time required for the accretion of the solar nebula. After a fragment separates from the interstellar cloud its central region is likely to be denser, and will collapse more rapidly, than the remainder of the fragment. A small solar nebula will therefore be formed at first when the central region ceases to collapse; that small nebula will grow by accretion of the remainder of the infalling fragment

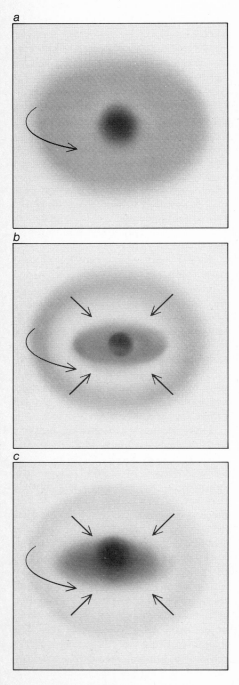

ACCRETION MODEL of the primitive solar nebula assumes that a central region of the cloud fragment collapses faster than the rest (a). It forms a small solar nebula: a central mass that is not yet the sun, surrounded by a disk of gas and dust grains, with more gas and dust concentrated around the periphery (b). The small nebula then grows by accretion over a long period of time (c).

over a period of time—probably between 10,000 and 100,000 years, a lot longer than the cooling and angular-momentum transport times we estimated.

These considerations have led me to a new picture of the primitive solar nebula that I am currently trying to define in detail. It is necessary to construct not a single model but an evolutionary sequence of models, beginning with a small solar nebula that grows through accretion over a time interval of perhaps 30,000 years. In this case the time for the redistribution of angular momentum remains short compared with the accretion time, so that much of the mass flows inward to form the sun not at the beginning of the accretion period but throughout that period; the mass of the nebula out in the region of planet formation remains a relatively small fraction of a solar mass throughout the period. As for cooling, the accreting gas is suddenly decelerated when it hits the surface of the solar nebula; the energy of its infall is converted into heat that is radiated from the surface. In the later stages of accretion that process keeps the surface layers in the region of planet formation at a temperature of perhaps a few hundred degrees; the temperature in the interior would be somewhat higher. And meanwhile the steady flow of mass toward the central spin axis diminishes dynamic instabilities within the nebula.

The two pictures of the primitive solar nebula, one derived from the masses of the planets and the other from the principles of star formation, thus seem to be converging to form an intermediate model of the initial solar nebula. In that model somewhat more than one solar mass has collected toward the spin axis but is not yet recognizable as the sun. It is surrounded by a disk of gas and dust amounting to perhaps a tenth of a solar mass. Farther out, beyond the region of planet formation, considerable additional amounts of mass are still falling toward the solar nebula.

The planets were created by the accumulation of interstellar grains and, in the case of the outer planets, the subsequent attraction and adherence of gases. The buildup of solid matter would have begun, I have recently calculated, in the collapsing gas cloud. Turbulent gas eddies would have accelerated the interstellar grains until they had large enough relative motions to begin to collide with one another. Having been formed out of material in stars and then ejected into interstellar space, where ices and other volatile constituents condensed on their

surface, the grains probably had a rather fluffy structure. It would not be surprising if such particles stuck to one another when they collided, forming clumps. As time passed the clumps of grains would collide with one another, sometimes amalgamating into larger clumps and sometimes breaking up into smaller ones. By the time the solar nebula had formed, many clumps were likely to have grown to a diameter measured in millimeters or centimeters.

Clumps of that size could settle through the gas toward the midplane of the nebula in tens or hundreds of years. Since their settling rate would vary with size there would be further collisions, increasing the size of the clumps and accelerating their fall toward the midplane. At that point, however, unless they were somehow able to grow substantially larger they would rapidly be lost to the inner solar nebula as a result of the gas-drag effect I mentioned above.

A critical process in planet formation may therefore be a mechanism recently proposed by Peter Goldreich of the California Institute of Technology and William R. Ward of Harvard University, which would give rise to those larger bodies. They showed that if there is a thin layer of condensed solids at the midplane of the nebula, with very little relative velocity among the particles, then a powerful gravitational instability mechanism will break up the thin sheet into bodies with diameters in the range of the diameters of asteroids: kilometers or tens of kilometers. The instability mechanism gradually operates over larger distances, attracting the asteroid-size bodies into loosely bound gravitating clusters of hundreds or thousands of bodies. The clusters remain unconsolidated because of the large angular momentum contained in their component bodies, which makes them rotate around common gravitating centers. When two clusters approach each other, however, they intermingle; the fluctuating gravitational field in the combined cluster leads to a violent dynamic relaxation of the motions of the bodies, so that many of them coalesce to form cores around which others go into orbit (although some of the bodies would be lost). The clusters interact with one another gravitationally over quite large distances; mutual perturbations gradually build up the velocities of the clusters with respect to one another, leading to further collisions that produce ever larger bodies.

The Goldreich-Ward instability mechanism would appear to be a powerful first step in the accumulation of planetary bodies. The subsequent stages in

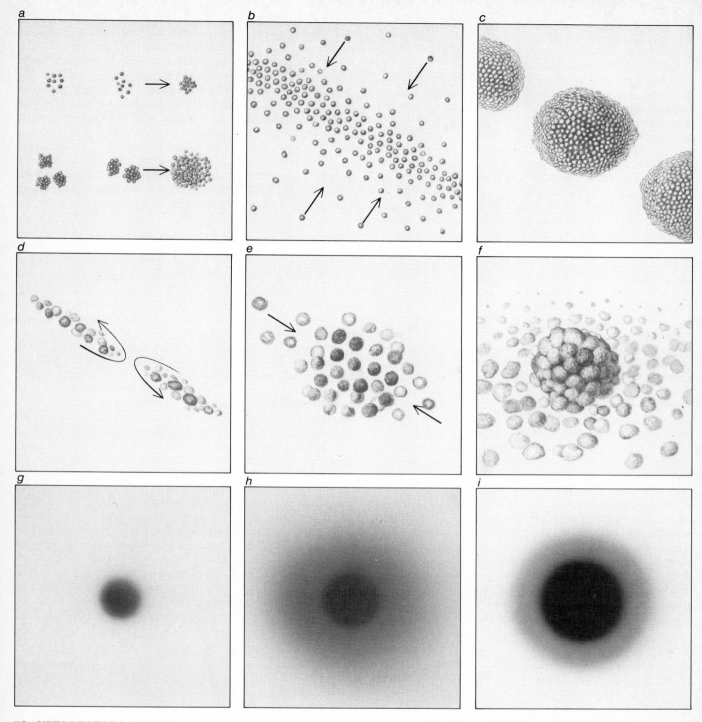

**PLANETS BEGIN TO FORM** when interstellar dust grains collide and stick to one another, forming ever larger clumps (*a*). The clumps fall toward the midplane of the nebula (*b*) and form a diffuse disk there. Gravitational instabilities collect this material into millions of bodies of asteroid size (*c*), which collect into gravitating clusters (*d*). When clusters collide and intermingle (*e*), their gravitational fields relax, and they coagulate into solid cores, perhaps with some bodies going into orbit around the cores (*f*). Continued accretion and consolidation may create a planet-size body (*g*). If the core gets larger, it may concentrate gas from the nebula gravitationally (*h*). A large enough core may make the gas collapse into a dense shell that constitutes most of the planet's mass (*i*).

the process are still highly speculative and were surely different in different regions of the solar nebula. The interstellar grains whose clumping initiates the accumulation process are those that have not been vaporized by the heat of the nebula; their materials, and therefore the materials of the larger bodies into which they are incorporated, would be different at different distances along the steep temperature gradient: metals, oxides and silicates in the region of the inner planets; similar rocky compounds and water ice farther out; rock, water ice and frozen methane and ammonia still farther out.

In the case of the smaller inner planets the progression to full size may be just a question of successive collisions and amalgamations of rocky bodies. In the case of the outer planets there are other considerations. Fausto Perri and I have recently considered the behavior of the primitive solar nebula as a large planetary core grows within it. As the mass of the core increases, gas in the solar nebula becomes gravitationally concentrated toward the core; with the continued growth of the core the amount of mass in the gas that is concentrated increases even more rapidly than the mass of the core itself. At some point the core reaches a critical size (which depends on the temperature conditions in the surrounding gas) such that the gas becomes hydrodynamically unstable and collapses onto the planetary core.

The major constituents of Jupiter and Saturn are the hydrogen and helium in their atmosphere, and we believe it was through this process of concentration and collapse that these planets acquired most of their mass. Hydrogen and helium account for a smaller fraction of the mass of Uranus and Neptune, probably indicating that their core never grew to the critical size for hydrodynamic collapse; those two planets did, however, grow large enough to retain much of the hydrogen and helium that was gravitationally concentrated toward their core. The inner planets, on the other hand, may be too small ever to have concentrated much of the nebular gas.

When the collapse events took place to form Jupiter and Saturn, local conservation of angular momentum in the gas would cause it to flatten into a disk around the planetary core. As time went on the two planets would sweep up essentially all the gas in their vicinity within the nebula. One can think of them as forming miniature versions of the primitive solar nebula: a central core of condensed rock and ice taking the place of the sun, with the gaseous disk around the core as the analogue of the solar nebula. Both of these large planets have systems of regular satellites incorporating considerable mass, which probably formed from a gaseous disk by processes quite analogous to the formation of the planets in the solar nebula.

Any theory of the origin and evolution of the solar system must account somehow for the comets, its most spectacular but least understood members. Jan Oort of the Leiden Observatory suggested some years ago that the comets inhabit an enormous volume of space centered on the sun, starting well beyond the outer planets and extending to a distance of perhaps 100,000 astronomical units. The total mass of the comets in this vast "Oort cloud" is probably equivalent to between one earth mass and 1,000 earth masses, which would account for between $10^{12}$ and $10^{15}$ comets. The comets we see are those few whose orbital elements are perturbed by a passing star in just the right way to send them plunging toward the center of the solar system.

A comet is a "dirty snowball," an aggregate of ice and rocky material, in the model first suggested by Fred L. Whipple of Harvard University. As a comet approaches the sun, gases are vaporized from it, accompanied by dust particles, to form the characteristic coma and tail. Analysis of the tails shows that the molecules that are vaporized are primarily water but also include exotic organic compounds. The dust, some of which comes to rest high in the earth's atmosphere, consists of fluffy clumps of fine-grained rocky material. A comet, in other words, is apparently an assembly of interstellar grains.

The comets must either have been made within the solar nebula and somehow ejected into the Oort cloud or else have been made out in the Oort cloud itself. Oort originally suggested that they were formed near Jupiter and perturbed by Jupiter's gravitational field into very large orbits that were subsequently rounded out by stellar perturbations. That would require the formation of a staggeringly large mass of comets, since many more would have been ejected from the solar system than were retained in the Oort cloud. Moreover, given the temperatures that must have prevailed near Jupiter it seems unlikely that molecules more complex than water would have been in solid form. More complex ices are possible farther out in the nebula, and so Whipple and others have suggested that the comets were formed in the neighborhood of Uranus and Neptune and sent out into the Oort cloud by the gravitational fields of those planets. That proposal, however, meets only one of the objections to the Oort hypothesis.

My own belief is that the comets were probably formed out in the Oort cloud itself. It is true that the collapsing gas of the fragment of cloud that became the primitive solar nebula was never dense enough so far from the center for the interstellar grains to have aggregated into sizable bodies out there. There is another possibility, however. Most of the stars in the galaxy are much less massive than the sun, suggesting that gas-cloud fragmentation sometimes continues at least down to fragments a tenth of a solar mass in size. Fragmentation may have gone further still. Small fragments could have been bound gravitationally to the primitive solar nebula in

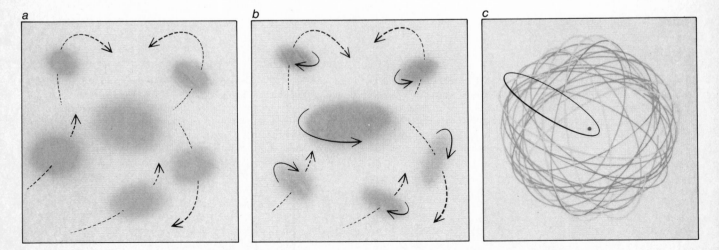

COMETS MAY HAVE FORMED from small cloud fragments that once were in orbit around the larger fragment that became the solar nebula (a). The small fragments spun down, like the solar one, to form disks in which comets were accumulated much as planets were (b). Eventually starlight could have evaporated the gases of these "cometary nebulas," leaving the comets in enormous orbits around the sun (c). From time to time a comet's orbit is perturbed by a passing star, and the new orbit brings it close to the sun.

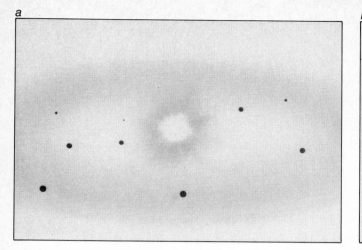

**SOLAR SYSTEM WAS CLEANED UP** by the "T Tauri wind." When gas contracting toward the center of the nebula reached a sufficient density, the nuclei of hydrogen atoms began to fuse and the sun began to shine (*a*). During its T Tauri phase the sun lost vast quantities of material. That material constituted an intense solar wind that could have blown away the remaining gas (*b*).

orbits traversing the region of the Oort cloud. Such fragments would form fairly large and very cool disks, ideal places for the comets to form. When the disks were ultimately heated by ultraviolet radiation from external stars, the gases would evaporate away and leave the comets in solar orbits [*see illustration on opposite page*].

After the planets had formed, much of the gas of the solar nebula must have remained in orbit around the sun, along with countless small bodies and large amounts of unconsolidated dust. There are only planets and asteroids in orbit now, with very little dust and almost no gas. How was the solar system cleaned up? As I mentioned above, young stars characteristically pass through what is called the T Tauri stage, when they eject matter at a prodigious rate: as much as one solar mass per million years! There is every reason to believe the sun passed through a similar phase, and the fierce "wind" of that ejected mass undoubtedly dissipated the solar nebula by carrying the residual gas off into space. That early solar wind would have stripped the inner planets of the remains of any primitive atmosphere of hydrogen and helium from the primitive solar nebula; the outer planets must have formed early enough to capture their hydrogen and helium before the solar wind began to blow. Moreover, if the accretion of gas from the original cloud fragment was still continuing, it would have been terminated by the wind, with the infalling gases being ejected back into interstellar space.

What determined when the T Tauri wind began? Why did it not blow away the primitive solar nebula long before the sun got so large? The thermonuclear reactions of hydrogen that constitute a stellar furnace are ignited only under extreme conditions of high temperature and high density. Perri and I have recently determined that the temperatures of the primitive solar nebula were so low that compressional heating of the gas could not have ignited the sun at a central density comparable to that of the sun today; the density would have had to be at least 100 times higher than it is now. Only then could the sun have adjusted itself into its present configuration, and only then could the thermonuclear furnace have been ignited and could the intense T Tauri wind begin to blow. An enormous amount of mass must have had to be gathered together to achieve such a density. Given the original temperature of the nebula, in other words, the sun had to reach a large mass before it could be a sun.

Once the sun had begun to shine and the T Tauri wind had blown away the gas, the stage was set for the final cleanup of interplanetary space and the completion of planet formation. The orbits of most of the small bodies in the solar system (other than the observed asteroids in their isolated belt) would have been continually modified by planetary perturbations. Over the course of a few hundred million years most such bodies either would have collided with one of the planets (the surfaces of Mercury, the moon and Mars still show the scars of that terminal bombardment) or would have been ejected from the solar system by a major planet, usually Jupiter.

The tiny dust particles were subjected to forces even stronger than gravitational perturbations: the effects of sunlight. The photons the particles absorb from the sun carry no angular momentum; the photons the particles radiate, however, carry off some of the angular momentum of the particles' orbital motion. The sunlight therefore acts as a resisting medium for the particles, making them spiral in toward the sun. Larger solids, up to a kilometer in diameter, are perturbed by sunlight in a different way. As such a body rotates the temperature of a section of its surface increases as long as it is on the sunlit side but decreases while it is on the dark side. One hemisphere of the body therefore emits considerably more radiation than the other. That gives rise to a preferential thrust that can perturb the orbit of the body either toward the sun or away from it, depending on the body's direction of rotation. Such bodies will eventually come close to one of the planets, whereupon they will be absorbed by collision or be ejected from the solar system. In these several ways the sunlight could have acted as a broom to sweep away much of the smaller debris left over from the formation of the solar system.

The account I have given makes a coherent, if incomplete, story. Many of its details remain highly speculative, however, and much of the story may have to be retold as new data test the present theories. In selecting and weaving together facts, ideas and hypotheses I have necessarily been strongly influenced by my own beliefs. The reader should be aware that others would weave quite different tapestries. These ancient questions are still far from being answered.

3

THE SUN

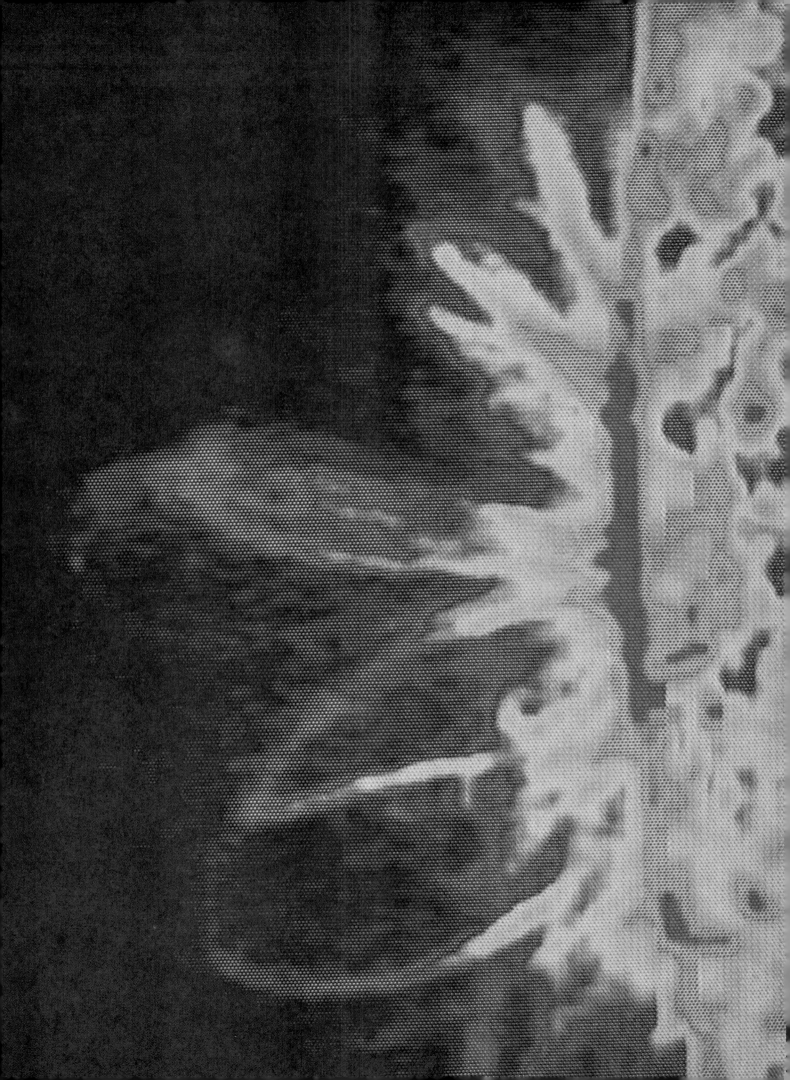

# The Sun

## E. N. PARKER

*Recent spacecraft observations have revealed spectacular new features of the solar surface and atmosphere. What happens inside the sun, however, has lately become more mysterious*

The earth has been warmed by the light of the sun for 4.6 billion years, and all life is maintained by the solar energy that is converted into chemical energy by plants. From the beginning of recorded history men recognized the life-giving role of the mysterious sun. In their awe of the burning disk they conceived it to be a deity, or to be under the direct care of one.

The sun is hardly less mysterious today. We possess a bewildering array of facts, gathered largely over the past 30 years in the outburst of technological development that has given birth to many new observational instruments. These facts make it clear that the sun has a complex nature that so far is only partly revealed.

The history of our scientific knowledge of the sun begins in the 18th and 19th centuries, when the properties of gases were thoroughly studied in the laboratory. From that knowledge, summarized in the law that the pressure of a gas is proportional both to its density and to its temperature, Jacob Robert Emden put forward the first crude theoretical model of the sun: a series of concentric gaseous shells. The two basic principles of the model were that at each level within the sun the internal pressure must be sufficient to support the weight of the overlying gas, and that the weight of the gas is determined by the gravitational attraction of all the gas that lies below it. Emden had no way of determining the temperature of the gas in the sun's interior, however, and so he could not draw any unique conclusions about the sun's mass and composition.

Also in the 19th century the temperature of the sun's surface was deduced from its brightness and from the distribution of that brightness with respect to the wavelengths of the visible spectrum. From the mean density of the sun that had already been deduced from gravitational theory, it was realized that the sun is a ball of hot hydrogen. Traces of heavier elements such as carbon, sodium, calcium and iron were identified by the spectroscope; indeed, helium was discovered by its spectral lines in the sun before it was isolated and identified in the laboratory by William Ramsay in 1895.

In the same period studies of terrestrial rocks indicated that life on the earth extended back millions of years, rather than the few thousand years derived from the interpretation of Scripture. Hence the sun must have been shining for at least that long. The enormous outpouring of energy from the sun ($10^{33}$ ergs per second) could not be due merely to the combustion of flammable material. Chemical fuels would have been exhausted in only a few thousand years. If the sun had contracted under the force of its own gravity at a rate of 100 feet per year, and was thus heated by the compression of its own gravitational field, it could have been shining for perhaps 30 million years. The geologist and the paleontologist, however, soon demanded more than that span. Their researches showed that the earth and the life on it were several hundred million years old at the least. Hence at the beginning of the 20th century it was clear that the sun had a source of internal energy far more potent than chemical combustion or gravitational contraction. New laws of physics were needed. To make the problem more complicated, it became clear that the earth had had a succession of profound climatic changes, the ice ages, which raised the question of whether the brightness of the sun varied.

With further knowledge came more mystery. It was gradually realized in the 19th century and the early 20th that there is violent activity on the surface of the sun. There are the sunspots, of course, but there are also the eruptions of the solar flares, which are associated with sunspots and were first identified in 1859 by the English solar observer Richard Carrington. An individual flare can last 30 minutes and emit a thousandth of the total energy of the sun from an area covering only a ten-thousandth of the total surface of the sun. Occasionally a flare is so bright that it is visible against the dazzling solar disk; it was such a flare that first caught Carrington's eye. Furthermore, spectacular auroras and "storms" in the earth's magnetic field seemed to be associated with solar flares.

The solar corona, the tenuous outer atmosphere of the sun, is visible during total eclipses of the sun and had been known for millenniums. It acquired a

ACTIVE LOOPS ON THE SUN are shown on the opposite page in a false-color picture made at the extreme-ultraviolet wavelength of 1,032 angstroms with the Harvard College Observatory ultraviolet spectroheliograph aboard the *Skylab* manned orbiting satellite. The loops, which are part of the inner corona, extend some 150,000 kilometers from the sun's western limb. Black and blue areas represent the least intense radiation, yellow and magenta the more intense and red the most intense. The intensity of the radiation is a function of both the temperature of the gas and its density. The loops are shown in the light of the emission from oxygen VI (oxygen atoms stripped of five electrons) at 300,000 degrees Kelvin. Thin red area along very edge of the solar disk is active region giving rise to loops.

new aspect in 1933, when it was discovered that its temperature was a million degrees Kelvin. The corona is visible because it scatters white light from the sun, just as dust particles in a sunbeam do. The observation that the gas in the corona has a temperature of a million degrees was startling and was immediately criticized as being absurd. How could the sun, with a surface temperature of 5,600 degrees K., heat a gas to a million degrees? The observation violated the second law of thermodynamics, that heat cannot flow continuously from a cold body to a hot one. Later Hannes Alfvén and Ludwig F. Biermann independently pointed out that the corona is heated by the dissipation of friction generated by convective motions below the visible surface of the sun. The temperature of the corona has nothing to do with the flow of heat from the visible surface, and so there is no thermodynamic limitation on the temperature of the corona. The

phenomenon is analogous to what happens when a man, with a body temperature of 37 degrees Celsius, rubs sticks together to achieve temperatures of several hundred degrees Celsius to light a fire. In principle there is no limit to the temperature that can be obtained in this way.

Some 15 years ago it was realized that the corona is continuously expanding outward into space. In doing so it creates a "wind" of charged particles moving at speeds of between 300 and 600 kilometers per second. It is the solar wind that is fundamentally responsible for the auroral and magnetic activity on the earth.

The concept of solar activity has been expanded in the past few years by observations from spacecraft. First of all, the expectation that active regions on the sun emit X rays was confirmed by direct observation, and the first X-ray

photographs of the sun were obtained by Herbert Friedman and his colleagues at the Naval Research Laboratory. The more recent X-ray photographs of the sun, made by Giuseppe Vaiana and his group at American Science and Engineering, Inc., show hundreds of tiny bright hot spots scattered over the surface of the sun. The individual spots wink on and off with a lifetime of about eight hours, many of them flaring up momentarily before they fade away. The recent spectacular observations from the manned orbiting observatory *Skylab* reveal the startling fact that such eruptions occur every few hours. There are many more eruptions than there are solar flares. Presumably it is these frequent eruptions, visible only from above the earth's atmosphere, that give rise to much of the variation in the solar wind and ultimately in the magnetic field of the earth.

How does one go about understanding such a complicated object as the sun? One can only start by asking the most primitive question of all: Apart from the complex solar eruptions, what is the basic nature of the sun's hot sphere?

With the advance of atomic physics and electromagnetic theory in the early part of this century it became possible to estimate the opacity of the solar material, that is, how transparent it is to the passage of heat and radiation. Moreover, it was clear that the temperature must increase inward through the sun with sufficient rapidity to give rise to the observed outward flow of heat. A. S. Eddington, Karl Schwarzschild, Subrahmanyan Chandrasekhar and others were then able to show that the sun's central temperature must be close to 10 million degrees K. and that its density must be nearly 100 times the density of water. The matter is gaseous rather than liquid or solid because at 10 million degrees the atoms are deprived of their electrons and there is no chemistry, or binding of one atom to another.

The energy source of the sun remained elusive throughout the first three decades of the century. In 1896 Henri Becquerel had discovered radioactivity, and the concept of subatomic particles appeared in the scientific literature. Einstein stated that mass and energy were equivalent, thereby implying that they were interchangeable, leading physicists such as Eddington to begin to speak of the direct conversion of matter into energy in the sun, although by processes still unknown. Perhaps it was possible that the electron and the proton annihilated each other at the tremendous pres-

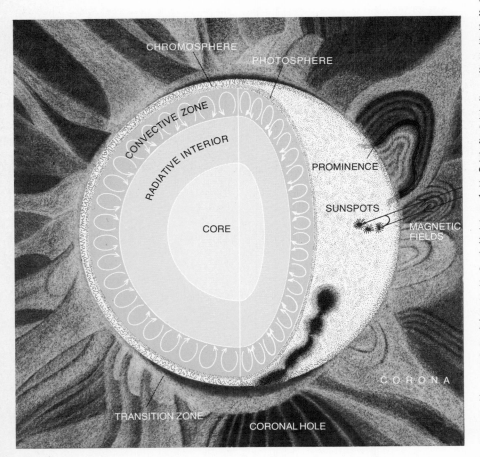

CROSS SECTION OF THE SUN shows the observable exterior features together with the hypothesized internal structure. Energy released by the proton-proton chain of thermonuclear reactions (*see illustration on pages 32 and 33*) in the core of the sun makes its way very gradually to the photosphere, or visible surface. It is transported by means of radiative processes, in which atoms absorb, reemit and scatter the radiation. At a point about 80 percent of the way from the core to the surface the gas becomes unstable to motions along the radius and the energy is then transported by convection. Shock waves from the convective zone carry energy up into the chromosphere through the transition zone and into the corona. Intertwined throughout structure of sun, from convective zone to the corona and beyond, are magnetic fields that give rise to sunspots, prominences and other activity.

sures inside the sun's core. The ideas were vague and spine-tingling.

It gradually came to be realized that the gas in the solar interior must consist only of free electrons and bare atomic nuclei (except for the rare atoms of heavier elements that manage to hang on to a few of their innermost electrons). The nuclei are mostly hydrogen, with an admixture of helium; only about 1 percent of them are carbon, nitrogen, oxygen and heavier elements. The bare nuclei collide with one another at an enormous rate. Since they are all positively charged, their collisions are generally cushioned by their mutual electrostatic repulsion, but nonetheless in the occasional more violent collisions the nuclei may get a "taste" of each other.

In 1919 Ernest Rutherford observed the first artificial transmutation of a nucleus. When he bombarded nitrogen nuclei with the helium nuclei emitted by radioactive bismuth, he found that once in a while a helium nucleus was absorbed by the nitrogen and only a single proton escaped, leaving behind a nucleus of oxygen. Rutherford, James Chadwick and then others began to explore the apparently endless variety of nuclear transformations that are possible when two nuclei come together with sufficient violence. In 1932 Chadwick found direct evidence that within the nucleus there are not only protons but also neutrons. In fact, within a dense nucleus an electron and a proton can be squeezed together to form a neutron. Thus the electron and the proton do not annihilate each other, as had been speculated earlier, but form a third particle.

With the advent of various machines that could accelerate nuclei to high energies it became possible to conduct studies of nuclear reactions under a wider variety of circumstances. The energy of the bombarding particle could be carefully controlled, and the likelihood of a transmutation could be accurately measured. Soon a peculiar effect was discovered: when a radioactive nucleus decayed by emitting an electron (the process known as beta decay), the energy given up by the nucleus was not entirely carried away by the electron. Sometimes the electron was ejected with the full amount of energy lost by the nucleus. Sometimes it was ejected with essentially no energy. Usually it was ejected at some intermediate energy. On the average the electron received only half of the energy that was lost by the nucleus. The other half apparently vanished. Did this mean that energy was not conserved, that nuclear forces were not

ERUPTIVE PROMINENCE of August 21, 1973, extended more than 400,000 kilometers above the surface of the sun. It was photographed in the light of singly ionized helium (helium II) by the Naval Research Laboratory ultraviolet spectroheliograph aboard the Apollo Telescope Mount on *Skylab* at the wavelength of 304 angstroms. On the scale of this picture the earth would be a dot somewhat larger than the period at the end of this sentence.

constrained by the principles that seemed to describe all other phenomena?

In 1933 Wolfgang Pauli asserted that the energy is conserved and that the only rational explanation is that the missing energy is carried away by an undiscovered particle. He deduced that the particle must be without charge and without mass except for its energy (that is, it has no rest mass), so that it could escape undetected. The hypothetical particle came to be known as the neutrino.

The big step in understanding the energy source of the sun was taken in 1939, when H. A. Bethe pointed out that there are two basic sequences of nuclear transmutations that can convert hydrogen into helium in the core of the sun. The transmutation of hydrogen to helium results in a net loss of .7 percent of the mass of the hydrogen nucleus and a

corresponding release of energy. One sequence is the carbon-nitrogen cycle, in which ultimately a carbon nucleus absorbs four protons and in the process emits two positrons (positively charged antiparticles of the electron) and becomes an unstable oxygen nucleus. The oxygen nucleus then splits into a carbon nucleus and a helium nucleus. Thus the original carbon nucleus is restored, and the net effect is the conversion of four protons into one helium nucleus.

The second sequence is the proton-proton chain, in which two protons collide and emit a positron and a neutrino to form deuterium, the heavy isotope of hydrogen that has a neutron as well as a proton in its nucleus. Another proton is then added to the deuterium to make the light isotope of helium, helium 3. Then two helium-3 nuclei combine to make one nucleus of ordinary helium,

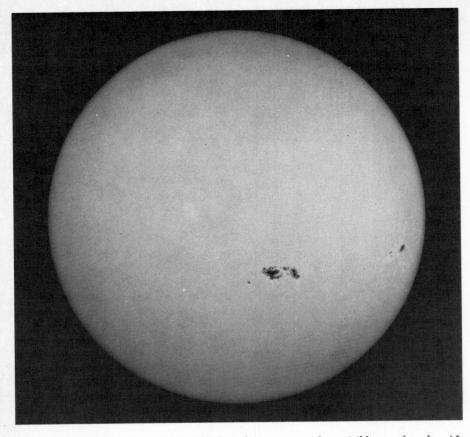

PHOTOGRAPH OF SUNSPOT GROUPS on the sun was made at visible wavelengths with McMath Solar Telescope at Kitt Peak National Observatory in Arizona on July 4, 1974, at minimum of sunspot cycle. Two sunspots of larger group are joined by a thin light bridge.

MAGNETOGRAM OF SUNSPOT GROUPS was made at the same time as the photograph above with the Solar Vacuum Telescope at Kitt Peak. White and black areas are regions of magnetic activity; white represents positive and black negative. The dark area in the central sunspot group corresponds to the light bridge in the photograph. The elongated appearance of the sun is an artifact of the way in which the magnetogram was made.

helium 4, and free two protons. The net result is again the conversion of four protons into one helium nucleus. The amount of energy that is liberated is approximately a million times greater than the energy released in a chemical reaction such as burning.

It now seemed that the mystery of the sun had been dispelled. So much energy is released by nuclear reactions that the sun has an expected lifetime of 10 billion years! It is about halfway through that lifetime at the moment. When the hydrogen is exhausted in its core in another five billion years, the core will contract until the temperature and the density are so high that helium nuclei will cleave together in groups of three to form carbon nuclei. The outer layers of the sun will puff outward past the orbit of Venus, and the sun will become a red-giant star. It will be a hot day on the earth when that happens.

Detailed studies over the past 20 years have shown that most of the stars in the sky can be understood in terms of the liberation of energy by the conversion of hydrogen to helium, with the carbon-nitrogen cycle being more important in stars hotter and more massive than the sun and the proton-proton chain being more important in stars about the size of the sun. With explanation and understanding, however, came another mystery. It was soon realized that the proton-proton chain can proceed in more than one way. For example, a helium-3 nucleus may occasionally combine with a nucleus of helium 4 rather than with another helium 3. The result is a light isotope of beryllium, beryllium 7, which in turn may react with a colliding proton to form boron 8. The boron 8 then decays, emitting a positron and a neutrino to form ordinary beryllium 8, which splits to yield two nuclei of helium 4.

This alternate branch of the proton-proton chain has no significant effect on the amount of energy released, because the result is still the conversion of hydrogen into helium. The important feature is that the decay of boron 8 involves the emission of a neutrino. The neutrino from that particular decay is emitted with so much energy that it is possible it can be detected on the earth. In particular, it has enough energy so that, if it collides with an atom of chlorine 37, it could transmute it into an atom of argon 37. The probability of the transformation is exceedingly small, but it was calculated that there should be so many neutrinos emitted from the sun that their detection should be feasible.

The task of building a neutrino detector was undertaken by Raymond Davis, Jr., of the Brookhaven National Laboratory. Detecting neutrinos is particularly difficult because all matter is almost entirely transparent to them: a neutrino could pass through a wall of lead 1,000 light-years thick and not be stopped. For this same reason neutrinos "shine" directly out of the core of the sun where the solar energy is generated. Davis' task was of heroic proportions, but the outcome would be a basic check—indeed, the only direct check—of our understanding of how the sun generates its energy. One mile underground in the Homestake Gold Mine in South Dakota (away from the nuclear reactions produced by cosmic rays) Davis set up a tank containing 100,000 gallons of the cleaning fluid perchloroethylene ($C_2Cl_4$). The argon that is transmuted from the chlorine by interaction with neutrinos is detected by its radioactivity.

The experiment has run for several years and has been continually refined and improved. Davis has shown that something is seriously wrong. There is no argon being produced in the tank, implying that there are no neutrinos being emitted from the sun. Davis' experiment shows that no more than about a tenth of the expected number of neutrinos are being detected.

What are we to conclude? On the one hand, can there be something wrong with the experiment? Davis is a careful worker who invites suggestions and criticism from his fellow scientists. He has introduced a number of refinements and extra tests to eliminate possible faults in his apparatus. On the other hand, is there some error in the calculated rate at which neutrinos are emitted from boron 8 in the sun? The expected production of neutrinos in the sun is very sensitive to the temperature in the solar interior. Is there something missing from the difficult calculation of the transfer of heat within the sun? Or is there something wrong with the fundamental physical principles that are applied to deduce the emission of neutrinos and the production of argon from chlorine?

There are many possibilities. Perhaps the luminosity of the sun goes up and down over a period of a million years or so. We recall the ice ages. At the present moment the core of the sun may be expanded and may be producing only a little energy and hence only a few neutrinos. If that is the explanation, it will have to be demonstrated from the basic principles of astronomy and physics. Or perhaps the deep interior of the sun, un-

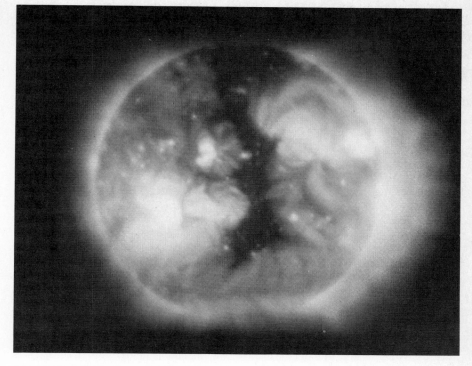

CORONAL HOLE, a large area where the temperature and density of the corona are reduced, was present on the sun on June 1, 1973. The picture, made with the X-ray telescope of American Science and Engineering, Inc., aboard *Skylab*, shows the corona as the bright indistinct regions. Coronal hole is large dark area in center of solar disk and is longer than 750,000 kilometers. Isolated white spots are X-ray bright points, regions of intense emission.

like the material at the surface, is free of the heavier elements, causing the gas to be more transparent at the core than is currently calculated and lowering the predicted temperature of the interior. With a reduced temperature the helium 3 will not combine with helium 4 to make the boron 8 that decays and emits the energetic neutrinos. Instead each helium-3 nucleus will combine only with another helium-3 nucleus to yield helium 4 and two protons.

The sensitivity of Davis' apparatus is being increased to the point where it soon should be able to detect the low-energy neutrinos emitted from the first step of the proton-proton chain, in which two protons collide and combine to form deuterium with the emission of a positron and a neutrino of somewhat lower energy than the neutrino emitted in the decay of boron 8. The lower-energy neutrinos are not as effective in making argon from chlorine, but they should be created in larger numbers. The production of deuterium from protons is the fundamental step in the generation of solar energy. If these neutrinos are also missing, the finding would indicate that the chain is not functioning at the present time—or that our understanding of how the sun generates its energy is faulty.

These difficulties should not cause us

to discard the hypothesis that the energy of the sun is produced by the conversion of hydrogen into helium. On the other hand, the discrepancy should not be swept under the rug. Although the best efforts of a number of very capable scientists have failed to clear up the difficulty, the problem was not even clearly defined until the past couple of years.

How the sun yields its energy is probably the most basic question we have to ask about the sun. If the present puzzle can be straightened out, the question will be answered in terms of nuclear reactions and other physical effects that are studied directly in the laboratory. The violent activity on the sun, however, presents a different problem. That activity seems to be the result of the flow of gas in magnetic fields. Although the basic physical properties and the basic mathematical equations pertaining to gas flow and magnetic fields are well known from experiments in the laboratory, the activity of the sun can take place only over dimensions of thousands of kilometers or more. The effects we see on the sun clearly cannot be duplicated and studied in the terrestrial laboratory.

Such a situation is not unknown in other fields. For example, weather is not something that can be studied in miniature in the laboratory. With its active

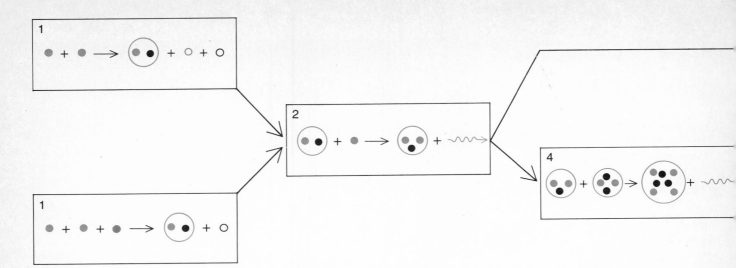

PROTON-PROTON CHAIN is the sequence of nuclear reactions for fusing hydrogen into helium that is believed to be the principal source of energy of the sun. In 99.75 percent of the reactions two protons combine to yield a nucleus of deuterium, or heavy hydrogen, which emits a positron and a neutrino (1). Occasionally two protons and an electron will combine to yield a nucleus of deuterium and a neutrino. The deuterium nucleus formed by either process then combines with a proton to form a nucleus of helium 3 (the light isotope of helium) and a photon (2). Some 86 percent of the time the helium-3 nucleus combines with another helium-3

fronts, jet streams and thunderstorms, it exists only over large areas. Therefore the study of solar activity is an interesting game. The observer endeavors to see what is happening on the sun, and then the theoretician strives to explain the observations from the basic principles of physics. The observations are made with telescopes on the ground, in high-altitude balloons and in space. Over the past three decades the ground-based visual telescopes have been developed into remarkably sophisticated and effective instruments for photographing the complex gas and magnetic configurations on the surface of the sun. Arrays of radio telescopes follow and record the outbursts of fast electrons from flares as they are ejected into space. Telescopes aboard spacecraft have observed the sun in the ultraviolet and X-ray regions of the spectrum, where the solar activity is particularly spectacular.

The magnetic fields in and around the sun do things that could not have been anticipated on the basis of laboratory experiments. The hot, electrically charged gases of the sun make an excellent conductor of electricity. At the center of the sun the gas is so hot that it conducts an electric current as well as copper does at room temperature. Elsewhere in the sun the gas conducts less well, but the diameter of the sun is so great that the body as a whole has a tremendous current-carrying capacity. The effect is that the magnetic fields in the sun are trapped in the gases and are transported and twisted as the gases stream up and down in the zone of convection just below the visible surface. The strength of

the magnetic fields around sunspots is typically 3,000 gauss (6,000 times the strength of the earth's field) across an area that may extend up to 50,000 kilometers. Fields of that strength can be generated by a laboratory electromagnet, but they are nonetheless impressive. One cannot hold on to an iron object in a magnetic field of 3,000 gauss: a screwdriver or a pair of pliers is wrenched from one's hand and slams into the nearest pole of the magnet.

There are quite a number of other exotic magnetic features on the sun. They range in scale from remarkable filigree structures at the limit of observation, with diameters of only a few hundred kilometers, to the spicules and bright spots in the X-ray region of the spectrum, to the eruptive prominences that sometimes burst thousands of kilometers up out of the sun. The upper layers of the solar atmosphere above the visible surface are profoundly affected by magnetic activity. The chromosphere (the dim irregular layer just above the visible surface) and the corona are enormously enhanced above active centers on the surface. One of the most remarkable discoveries made with the instruments aboard *Skylab* was the holes in the corona over the poles of the sun and over other regions without visible activity. It appears, paradoxically, that the coronal holes are the source of the high-speed ionized particles of the solar wind. It was known, but not seriously appreciated or understood before *Skylab*, that magnetic activity on the earth is sometimes actually depressed

when we face directly into active regions on the sun. The effect was obscured by the fact that the flares associated with the active regions did sometimes induce strong activity on the earth. In the absence of such large flares magnetic agitation and auroral displays are sometimes less conspicuous when an active region on the sun is facing us.

The magnetic fields in the sun show a curious tendency to concentrate themselves into isolated tubes, which often react violently with one another when they meet. Their behavior is still not understood in spite of a number of accepted "explanations" that have appeared in the scientific literature. I shall restrict my own exposition to the behavior of a conspicuous sunspot, which will suffice to give a glimpse of the perplexing and urgent problems that confront us.

A sunspot is a broad, shallow depression whose surface lies a few hundred kilometers below the overall visible surface of the sun. The umbra, or central floor of the depression, appears black in photographs because it is less than a fourth as bright as the adjacent surface. The surface temperature of the umbra is only 3,900 degrees K. compared with the normal 5,600 degrees of the rest of the surface. The umbra is a cool, rarefied area on the surface, and it is evidently the pressure of the surrounding surface that compresses and confines the intense magnetic field associated with the sunspot. It is widely believed that the intense magnetic field of the sunspot is responsible for the reduced temperature of the umbra, but the con-

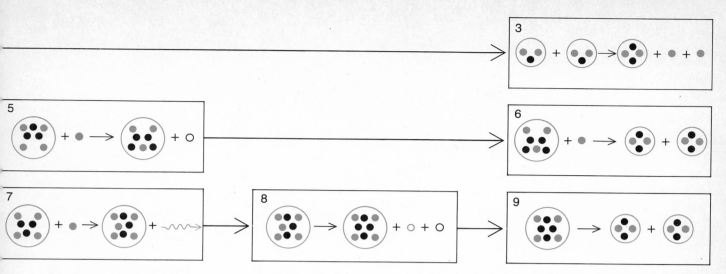

nucleus, forming a nucleus of ordinary helium 4 and liberating the two protons (3). There is an alternate branch of the proton-proton chain whereby the helium-3 nucleus combines with a nucleus of helium 4 to form a nucleus of beryllium 7 and a photon (4). Almost all the time the beryllium-7 nucleus then absorbs an electron and is transmuted into a nucleus of lithium 7 and a neutrino (5).

The lithium-7 nucleus next combines with a proton and splits into two nuclei of helium 4 (6). Rarely the beryllium-7 nucleus absorbs a proton to become a nucleus of boron 8 plus a photon (7). Boron-8 nucleus then decays into a nucleus of beryllium 8, a positron and a neutrino (8), which should be detectable on the earth. Finally, the beryllium-8 nucleus splits apart into two nuclei of helium 4 (9).

nection has yet to be fully demonstrated.

The umbra is surrounded by the penumbra, a striated gray border that slopes up and out from the umbra to the surrounding surface. The magnetic field of the sunspot emerges through the umbra and spreads out across the penumbra. Beyond the penumbra the magnetic field reenters the sun, often in a neighboring sunspot of opposite magnetic polarity. In some cases there is a region, known as a moat, immediately outside the sunspot that is remarkably free of magnetic fields. Evidently the magnetic field arches up over the moat before it reenters the surface at some distance from the spot.

A decade ago Jacques M. Beckers of the Sacramento Peak Observatory and E. H. Schröter, who is now at the University of Göttingen, observed that tiny clumps of concentrated magnetic field move into and out of the edge of sunspots, apparently participating in the growth or the decay of the spot. Sunspots often appear in a group with a complex structure that consists of many individual spots of opposite polarity mixed in complex patterns. It is in these complex and rapidly changing sunspot groups that one is most likely to see big solar flares. Possibly the flares are a consequence of the entangling of magnetic fields of opposite polarity that annihilate each other. A large flare may liberate $10^{32}$ ergs, equivalent to the amount of energy that would be released by the annihilation of a magnetic field of 1,000 gauss extending through a volume measuring 15,000 kilometers on a side. The largest sunspots have a field of some 4,000 gauss and a diameter of as much as 50,000 kilometers. The largest flares have bright filaments extending for 100,000 kilometers across the solar disk. Recent X-ray observations show that in each flare there is a central hot region, presumably the scene of the main action, that is much more concentrated than the bright filaments and occupies an area less than 10,000 kilometers across.

Sunspots are found in bipolar groups with their magnetic field emerging from the spots on the eastern side and reentering through the spots on the western side (or vice versa). The bipolar groups in the northern hemisphere of the sun are opposite in this respect to the groups in the southern hemisphere, and both reverse their polarity with each 11-year half of the sunspot cycle. The bipolar sunspot groups indicate that below the surface of the sun there are strong magnetic fields running east and west that are of opposite polarity in the northern and southern hemispheres.

The individual sunspots of a group usually last for only a few weeks. The first spots of a new sunspot cycle appear in a broad band centered on a latitude of 40 degrees in both the northern hemisphere and the southern. As the cycle progresses the two regions of sunspot formation migrate toward the equator, arriving between 10 and 14 years later. There they die out at about the same time that the new spots of the next sunspot cycle are appearing at middle latitudes. The cycle repeats itself with the new sunspots of reversed polarity.

The earth moves around the sun on an orbit that is inclined at an angle of about seven degrees to the solar equatorial plane, so that the sunspots and flares point radially toward the earth only in the last years of the sunspot cycle. Unfortunately for the physicist who studies solar activity the time of greatest activity is when the sunspots are forming at 15 or 20 degrees from the equator. The big blasts of the solar wind from the large number of flares at a solar latitude of 20 degrees can be detected in the way they affect radio signals passing through the wind from distant celestial radio sources.

In contrast, the solar wind from the poles of the sun seems to be relatively steady and very swift: about 50 percent faster than the solar wind in the plane of the sun's equator. The reason the solar wind is so fast at the sun's poles is evidently that the polar regions are coronal holes. Firsthand information on the strength of the blasts and on the solar wind from the poles, however, can be obtained only when we send a spacecraft far out of the plane of the orbit of the earth. One could accomplish such a mission by launching a spacecraft on a trajectory to Jupiter and then letting the planet's combined speed and gravitational field swing the craft around and up over the poles of the sun.

Sunspots have behaved strangely in past centuries. They were discovered in the Western world in 1611 by Johannes Fabricius and Galileo following the invention of the telescope. (Chinese observers, using only the unaided eye, had known and recorded sunspots for at least 1,000 years before.) The 11-year sunspot cycle, however, was not recognized until late in the 18th century, and it was not confirmed until 1843 by Heinrich Schwabe. The curiously slow recognition of this rather obvious phenomenon may be understood at least in part by the behavior of the sun itself.

In 1895 and again in 1922 E. W. Maunder published articles calling attention to the strange behavior of the sun in the 17th century. The old records of sunspots indicate that there were two maximums in the sunspot cycle in the 30 years following the discovery of sunspots in 1611. The maximums were about 15 years apart. Then the records show that the number of sunspots and the associated solar activity declined to a very low level about 1645 and remained almost entirely absent until 1715. After 1715 the sunspot cycle as we know it today appeared and has continued ever since. The significance of the prolonged absence of sunspots could not be appreciated at the time because there had been 34 years with sunspots

and 70 years without them. Who was to say what was normal? The 260 years of sunspot activity since 1715 has introduced the prejudice that the 11-year sunspot cycle with its thousands of spots in each cycle is the norm. We can only wait to see what the next few centuries bring.

During the 70 years of inactivity observers often had to wait years to see a single sunspot, whereas now there are usually a few spots showing even during the minimum of the sunspot cycle. The importance of the absence of sunspots lies not in the sunspots themselves but in the absence of the solar activity associated with them. The solar corona, which is heated in large degree by the friction and agitation of the active regions on the sun, was not seen during the solar eclipses of the 70-year period of inactivity. The entire corona seems to have been one big coronal hole.

The magnetic storms and auroral displays that are normally common in the British Isles and in the Scandinavian

countries virtually disappeared during the 70-year period, so that in 1715, when solar activity reappeared, the auroral displays caused wonder and consternation in such places as Copenhagen and Stockholm. Perhaps the most curious and important effect was discovered by A. E. Douglass in the course of his pioneering work on tree-ring dating, which was first published about 1920. Douglass noticed a general cyclical variation in the annual growth of trees: there was a tendency for the annual growth rate to increase and then decrease over each decade or two. As his work progressed and he undertook to study samples of wood several centuries old, he observed that the last half of the 17th century was remarkable for the absence of the usual cyclical variation. The tree rings varied little in width, and they varied without any evident pattern, making it difficult to match the annual rings of one tree with those of another and thereby establish a reliable sequence of ring dates. Douglass was at a loss to account for the effect, other than to attribute it to some aspect of the environment that was more uniform over that period, causing the growth of trees to be more regular.

It was therefore with considerable interest that in 1922 Douglass read Maunder's paper in *Journal of the British Astronomical Association* reporting the absence of the sunspot cycle and solar activity over just that period of time when there was so much trouble with the tree-ring dating. Douglass wrote to Maunder, who reported the tree-ring effect to the British Astronomical Association.

Why terrestrial weather and plant growth should vary in coincidence with solar activity is a mystery that has been unresolved in the half-century since Douglass' observation. Fortunately for those interested in the matter there is now renewed curiosity about possible connections between solar activity and the terrestrial environment. The problem is difficult and racked by heated controversy, but it is too important to be ignored. The growth of plants, after all, is the basis for life on the earth.

Altogether the sun has proved to be more of a mystery than we might have wished. We have come a long way in understanding it, but we still have a long way to go. As we strive to understand the present mysteries we are sure to uncover others. Indeed, it would be a pity if it were otherwise. The riddles the sun presents are signposts to new horizons.

NEUTRINO DETECTOR was built by Raymond Davis, Jr., of the Brookhaven National Laboratory in an attempt to detect the neutrinos that were expected from the decay of boron 8 in the proton-proton chain. The detector is a tank containing 100,000 gallons of the dry-cleaning solvent perchloroethylene ($C_2Cl_4$) buried under a mile of rock in the Homestake Gold Mine near Lead, S.D. The solvent is 85 percent chlorine 37. When a neutrino is absorbed by a nucleus of chlorine 37, a nucleus of the radioactive isotope argon 37 is formed. The tank is swept with helium gas every two or three months to remove the argon 37, which is then detected by its radioactive decay. Since 1968, when the experiment was begun, less than 3 percent of the predicted amount of argon 37 has been detected, implying that far fewer neutrinos are being produced in the sun's interior than is predicted by theory.

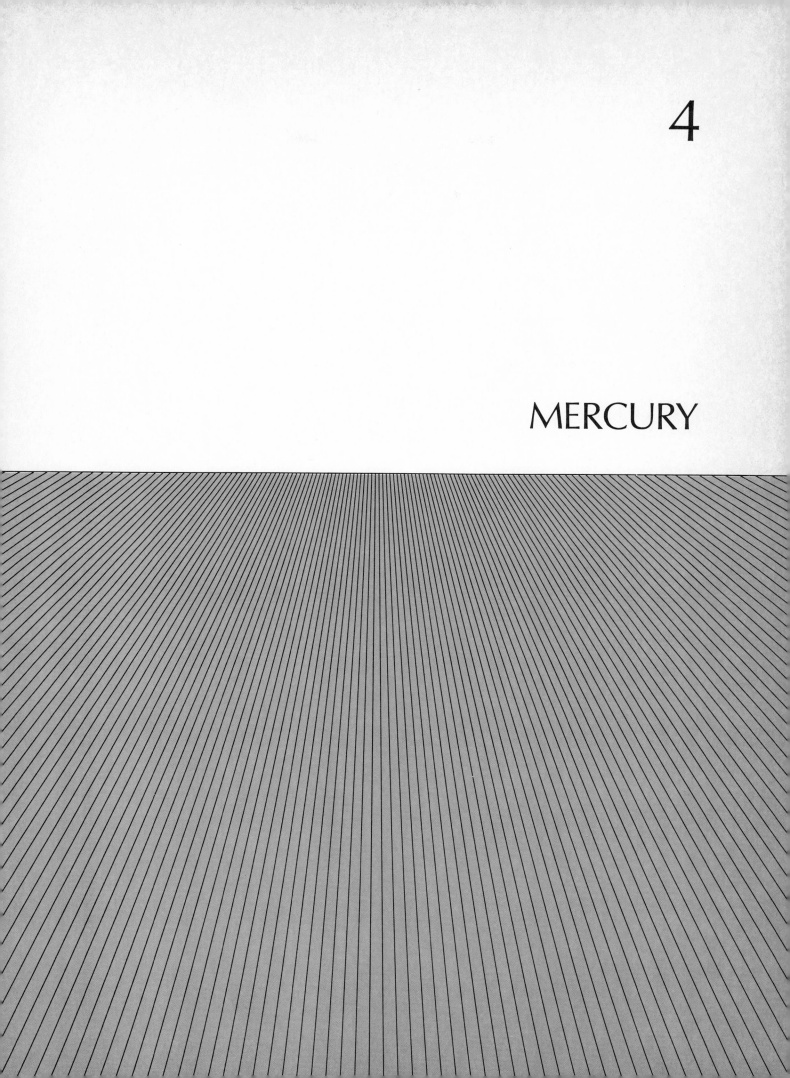

4

MERCURY

# Mercury

BRUCE C. MURRAY

*The remarkable pictures made by the spacecraft
Mariner 10 have revealed a planetary paradox:
Although Mercury is like the earth on
the inside, it is like the moon on the outside*

There is a story that as Copernicus lay on his deathbed he lamented never having seen the planet Mercury. The story seems implausible because in northern Europe the planet is occasionally visible at twilight. Even if Copernicus could have viewed Mercury through a modern telescope, however, he would have been presented with a singularly unrewarding image. Only a few vague markings can be discerned through a telescope. They are so faint that optical astronomers were long misled into assigning the planet an incorrect rate of rotation.

Last year, 501 years after the founder of modern astronomy was born, the spacecraft *Mariner 10* passed within a few hundred kilometers of Mercury, providing both a 5,000-fold increase in photographic resolution of the planet's surface features and entirely new measurements of phenomena in the planet's immediate environment. Suddenly Mercury was plucked from obscurity and placed in an observational status comparable to that of the moon before the modern age of space exploration. Furthermore, *Mariner 10* is in an orbit that carries it back to the vicinity of Mercury every 176 days. As a result the spacecraft transmitted a second set of close-up photographs on September 21, 1974, 176 days after its first encounter with the planet, and an extremely valuable third set of observations, including a limited set of high-resolution pictures, on March 16 of this year, shortly before it exhausted its supply of gas for stabilizing its attitude in space.

Before *Mariner 10*'s highly successful voyage it was recognized that Mercury is covered with at least a thin layer of finely divided dark silicate material very similar to that on the moon. Allowing for the difference in distance from the sun, Mercury closely mimics the moon in the way it reflects sunlight and radar pulses and in its emission of infrared radiation and radio waves. Yet it has been known on the basis of Mercury's size and mass that the planet is much denser than the moon or Mars and only slightly less dense than the earth. The earth's bulk density is greater than the laboratory density of its constituent materials, since much of the earth's substance is under high pressure in its interior. Hence the high bulk density of the much less massive Mercury implies that Mercury contains an even greater abundance than the earth of heavy elements, particularly iron.

Even the two most elementary facts about Mercury inferred from ground-based observations—the nature of its surface materials and its density—raised difficult questions. Could Mercury be composed of a homogeneous mixture of iron and silicate materials, as is the case with certain kinds of meteorites? Alternatively, could Mercury have a large earthlike iron core enclosed by a relatively thin silicate mantle and crust? If Mercury is chemically differentiated as the earth is, and the evidence now points in that direction, the diameter of its iron core is fully three-fourths the diameter of the planet, or the size of the moon!

Since Mercury is the innermost planet in the solar system, never more than 28 angular degrees from the sun in the sky, it is notoriously difficult to study by conventional astronomical techniques. We now appreciate how seriously these techniques can be compromised by contamination with sunlight. (Mars, in comparison, presents its largest image in the middle of the night, when the earth lies directly between it and the sun.) As recently as 1962 a leading expert in the visual and photographic observation of the planets wrote in what was then the most authoritative book on planetary astronomy that Mercury rotates at a rate such that one hemisphere constantly faces the sun, as one hemisphere of the moon constantly faces the earth. This synchronous rotation meant that Mercury supposedly turned on its axis once every 88 days, in step with the period of its revolution around the sun. Indeed, the planet was said to be synchronous with the sun to within one part in 10,-000. In the same period spectroscopists working at two different wavelengths concluded that Mercury has a thin atmosphere. That conclusion received independent support from purported variations across the disk of the planet in the degree to which reflected sunlight

CRATERED SURFACE OF MERCURY was photographed for the first time late in March of last year by cameras aboard *Mariner 10*. This high-resolution picture shows a typical heavily cratered region, strongly resembling the surface of the moon, on the equator of the planet. The pictures returned by *Mariner 10* made it possible to relate a previously agreed-on longitude system for Mercury to a specific topographical feature for the purposes of detailed mapping. In 1970 the International Astronomical Union had defined the origin of planetographic longitudes as the meridian crossing the subsolar point of the first perihelion passage of 1950. It has now been agreed that the 20-degree meridian of Mercury passes through the center of a particular small crater immediately adjacent to the large crater near the center of this picture. The small crater, which is 1.5 kilometers in diameter and lies .58 degree south of the equator, is at the foot of the outer rim of the large crater at a position equivalent to eight o'clock on a clock dial. It has been named Hun Kal, which stands for 20 in the language of the Maya, who used a base-20 number system. The photograph is reproduced with north at right in order to include as much of *Mariner 10* frame as possible.

was plane-polarized. We now know that Mercury has no atmosphere whatever, and has not had one for billions of years. And the planet does not revolve synchronously around the sun.

On the other hand, imaginative artists had commonly depicted the surface of Mercury as resembling the surface of the moon, and their intuition has proved to be correct. The *Mariner 10* photographs reveal that Mercury's surface is remarkably similar to the moon's, not only in its features but also in the sequence of events that was required to produce them. Mercury thus presents something of a planetary paradox: it is like the moon on the outside, yet like the earth on the inside, even to exhibiting an earthlike magnetic field.

In 1962, after centuries of unsatisfactory observations at visible wavelengths, radio waves from Mercury were detected. Radio astronomers from the University of Michigan observed the planet near elongation, when half of its disk, as it is seen from the earth, is in sunlight and half is in shadow. If Mercury were in synchronous rotation around the sun, the dark side would never receive any direct radiation from the sun and would be perpetually cold. Hence the thermal emission (including radio waves) from the dark side should be extremely low, well below the limit of detection. The Michigan workers were surprised to discover a substantial total radio flux evidently arising from both the dark and the sunlit halves of the planet, corresponding to an overall average near-surface temperature of 350 to 400 degrees Kelvin (170 to 260 degrees Fahrenheit). Such an average apparent temperature is exactly what any moonlike object would exhibit at the orbit of Mercury if it were rotating on its axis more rapidly than the once-a-revolution synchronous rate. Astronomers were so committed to the idea of synchronous rotation, however, that it was generally presumed that the anom-

**TWO HEMISPHERES OF MERCURY,** each approximately half in shadow, were photographed during *Mariner 10*'s first encounter with the planet in March of last year. The mosaic of high-resolution pictures at the left shows the "incoming" hemisphere: the hemisphere visible as the spacecraft approached the planet, before sweeping behind its dark side. The evening terminator, the shadow line at the right, lies near 10 degrees west longitude. Since *Mariner 10* approached Mercury from below the plane of its orbit, the center of the disk in the view at the left is about 20 degrees south of the equator. The area within the upper rectangle, which encloses the bright-rayed Kuiper Crater, is shown enlarged in the illustrations on the opposite page. The area within the lower rectangle appears enlarged on page 40. The mosaic at the right shows the "outgoing" hemisphere of Mercury, the hemisphere visible as *Mariner 10* "looked back" after passing behind the dark side of the planet. The spacecraft is now viewing Mercury from a point about 20 degrees north of the equator. The morning terminator, the shadow line at the left, lies near 190 degrees west longitude. The large impact structure named the Caloris Basin, comparable to the Imbrium Basin on the moon, is half-visible on the terminator just north of the center of the disk. The region inside the rectangle is shown on pages 42 and 43 in a sequence of pictures of increasing resolution.

alous thermal emission from the dark half must indicate the presence of an atmosphere capable of transporting heat from the day side to the night side.

In 1965 Rolf B. Dyce and Gordon H. Pettengill carefully measured the differences in frequency among returning radar pulses beamed at the edges of Mercury from the Arecibo Observatory. They concluded that the planet did not rotate synchronously around the sun but instead had a rotation period of $59 \pm 5$ days in the direct sense (in the same sense as the earth's rotation). They did not, however, mention in the scientific publication of their results that this finding would explain the "anomalous" heat emission from the dark side of the planet. Even a year later a comprehensive review article was devoted to the putative atmosphere of Mercury and its presumed role in the transport of heat.

Why a 59-day rotation period? Giuseppe Colombo, an Italian dynamicist with a long interest in Mercury, quickly recognized that a period of 59 days stood in relation to the 88-day period of the Mercurian year about in the ratio 2 : 3. Colombo conjectured that Mercury's rotation period is, in fact, precisely 58.65 days, which means that the planet would rotate exactly three times while circling the sun twice, thereby exhibiting the phenomenon of spin-orbit coupling. The conjecture has been fully confirmed not only by further radar observations but also by the photographs from *Mariner 10*.

It is extremely improbable that Mercury exhibits spin-orbit coupling simply by coincidence. It is more likely that tidal interaction with the sun has removed angular momentum and slowed the planet sufficiently from a higher original rate of spin to trap it into the present resonant period. Such a theory was quickly developed by Peter Goldreich and Stanton J. Peale and by Colombo and Irwin I. Shapiro.

The success of the *Mariner 10* mission depended on a three-body interaction calling for an assist from Venus. The spacecraft initially followed a course that took it close to the intervening planet, where its trajectory was perturbed toward the orbit of Mercury. In this way an entirely different orbit around the sun was achieved, an orbit otherwise quite unobtainable with a launch vehicle of the Mariner class. The new orbit carried *Mariner 10*, nearly at perihelion (the point closest to the sun), to an encounter with Mercury when the planet, traveling in its own eccentric orbit, was nearly at aphelion (the point farthest from the sun). An initially unforeseen property of

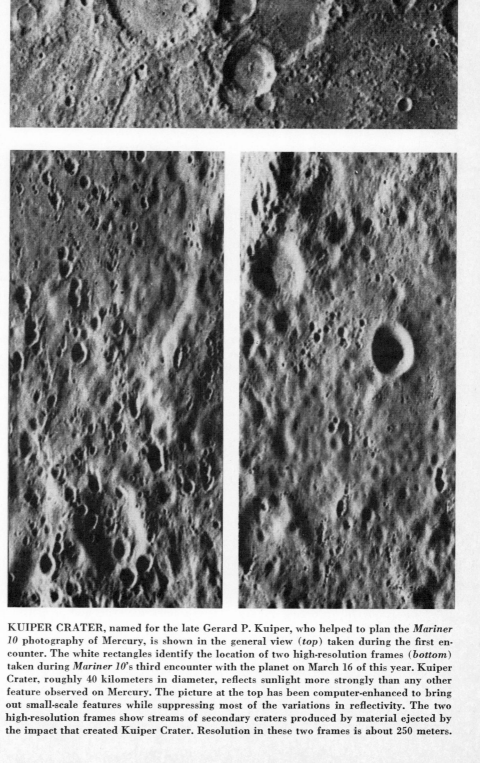

**KUIPER CRATER**, named for the late Gerard P. Kuiper, who helped to plan the *Mariner 10* photography of Mercury, is shown in the general view (*top*) taken during the first encounter. The white rectangles identify the location of two high-resolution frames (*bottom*) taken during *Mariner 10*'s third encounter with the planet on March 16 of this year. Kuiper Crater, roughly 40 kilometers in diameter, reflects sunlight more strongly than any other feature observed on Mercury. The picture at the top has been computer-enhanced to bring out small-scale features while suppressing most of the variations in reflectivity. The two high-resolution frames show streams of secondary craters produced by material ejected by the impact that created Kuiper Crater. Resolution in these two frames is about 250 meters.

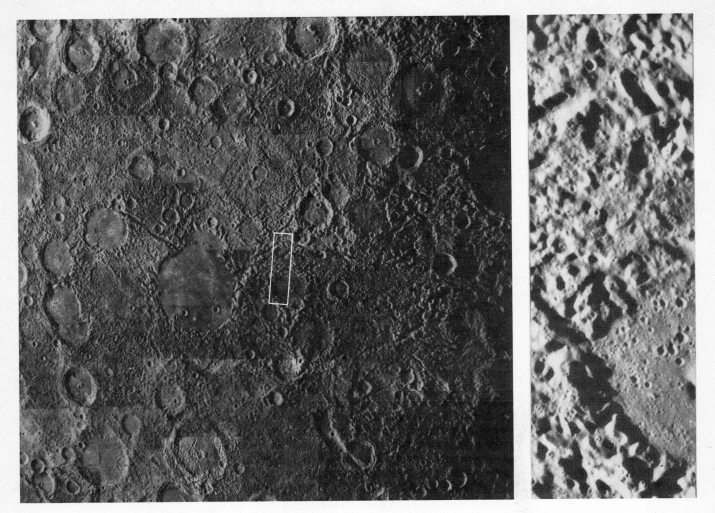

**PECULIAR TERRAIN** lies southeast of Kuiper Crater and antipodal to the Caloris Basin. The area shown in the picture at the left lies within the lower rectangle in the mosaic at the left on page 38. One hypothesis is that the hilly and lineated terrain shown here was created by the seismic effects of the great impact that created the Caloris Basin on the opposite side of the planet. The large cra- ter near the center of the picture, with two small craters in its floor, is about 170 kilometers in diameter. The area within the white rectangle is shown in the high-resolution picture at the right, which was taken during *Mariner 10*'s third encounter with the planet. The picture resolves surface features as small as 450 meters. The half-visible large crater is approximately 55 kilometers in diameter.

this exquisite celestial billiard shot is that *Mariner 10*'s new orbit has a period exactly twice that of Mercury's. As a result *Mariner 10* will continue to return every two Mercurian years to pass Mercury at exactly the same heliocentric longitude. Since Mercury itself spins precisely three times on its own axis in two Mercurian years, the spatial orientation and surface illumination of the planet at each encounter with *Mariner 10* will be exactly the same [*see top illustration on page 44*]. Thus Mercury, the sun and the spacecraft are dynamically in a state of triple resonance.

Mercury has the harshest surface environment of any planet in the solar system. When Mercury is at perihelion, it receives 10 times as much solar energy per unit of surface area as the moon. Noontime temperatures at the equator of Mercury soar to 700 degrees K., and in the dark hemisphere the surface cools radiatively to less than 100 degrees. Furthermore, "noontime" lasts a long time at Mercury's perihelion because of the coupling of the planet's rotation with the period of its eccentric orbit. An observer on Mercury at perihelion would see the sun slow to a complete halt in its motion across the sky and then move slightly in a retrograde direction (westward through the constellations) for eight days [*see bottom illustration on page 44*]. The reason is that the orbital angular velocity exceeds the spin angular velocity near perihelion. Moreover, the areas subjected to the longer period of solar radiation at perihelion are always near the same longitudes: 0 degrees and 180 degrees. The longitudes 90 degrees and 270 degrees receive their maximum solar irradiation at aphelion. As a result the 0-degree and 180-degree meridians receive two and a half times more solar radiation overall than the meridians 90 degrees away from them. Hence even though Mercury's spin axis is probably perpendicular to the plane of the planet's orbit, so that there would be no seasonal variations with latitude as there are on the earth and on Mars, the spin-orbit coupling of Mercury gives rise to a seasonal variation in temperature with longitude.

Another interesting property of Mercury is that in its equatorial regions the subsurface temperatures are always above the freezing point of water and in the polar regions the subsurface temperatures are well below freezing. In contrast, liquid water of internal origin cannot reach the surface of the moon or Mars (except through volcanic activity) because the subsurface temperatures are everywhere below freezing for a depth of many kilometers. In view of the large longitudinal variation in the influx of solar radiation and the potential for a latitudinal variation in chemical weathering conceivably associated with the occasional release of subsurface water, one might expect to perceive some characteristic effects on the surface at appropriate geographic locations on the planet. Actually we have been surprised to find no such effects either in radar maps made from the earth or in the photographs sent back by *Mariner 10*.

The illuminated surface observed by *Mariner 10* as it first approached Mercury is dominated by craters and basins, creating a landscape that could easily be mistaken for parts of the moon. There are, however, significant differences. The heavily cratered regions of Mercury exhibit conspicuous plains, or relatively smooth areas, between the craters and the basins, whereas the highlands of the moon generally show densely packed and overlapping craters. The intercrater plains appear in many cases to predate the formation of most of Mercury's large impact craters. The surface of Mercury is also unlike the surface of the moon in that it is not saturated with large craters having diameters between 20 and 50 kilometers.

One factor contributing to the difference in surface appearance between Mercury and the moon has been suggested by Donald E. Gault of the National Aeronautics and Space Administration. He points out that since on the surface of Mercury the force of gravity is twice that on the moon, material ejected from a primary impact crater on Mercury will cover an area only a sixth as large as the area covered on the moon for an impact crater of the same size. On Mercury secondary impact craters are much more closely clustered around primary craters than they are on the moon. As a result the topographic record of early events may be better preserved on Mercury than it is on the moon, where ejecta from the most recent impact basins has blanketed much of the earlier surface.

Another important difference between the heavily cratered regions of Mercury and those of the moon is the ubiquitous presence on Mercury of shallowly scalloped cliffs running for hundreds of kilometers. The structure of these features, termed lobate scarps, suggests they resulted from an early period of crustal shortening on a global scale. Such features are not conspicuous on the moon or on Mars. On the contrary, on those two lower-density bodies one sees tectonic evidence of crustal stretching. Robert G. Strom of the University of

**VIEWS TAKEN 176 DAYS APART** by *Mariner 10* on its first encounter (*top*) and on its second encounter (*bottom*) show that the angle of solar illumination in each case is virtually identical, thereby supporting earlier evidence that Mercury rotates on its axis exactly three times while going around the sun twice. The planet's orbital period is 88 days and its rotation period is 58.65 days. Bottom picture shows a smaller area and is more distorted than top picture because it was taken closer to the planet and at a more oblique angle.

Arizona and others have speculated that the lobate scarps of Mercury are the result of a long period of crustal shortening produced by the slow cooling and contraction of Mercury's large iron core. In any case the very existence of large, well-preserved craters on Mercury, which are probably three or four billion years old, is evidence that there has been no planetwide melting or earthlike migration of crustal plates since that time. Furthermore, the evident lack of surface erosion rules out any tangible atmosphere for Mercury as far back as the time when the craters were made. In contrast the surface of Mars illustrates clearly how even a tenuous atmosphere quickly blankets and modifies the appearance of large craters, notably removing the initially conspicuous bright "rays" of ejecta that radiate from them.

While the television cameras of *Mariner 10* were revealing the ancient cratered terrain of Mercury, the magnetometer, the plasma probe and the charged-particle detector aboard the spacecraft were recording an interaction with the "wind" of charged particles from the sun that was much stronger than had been anticipated. The track of *Mariner 10* had been calculated so that the spacecraft would pass the dark side of the planet on its initial encounter; in that way the instruments would be able to investigate the wake left by Mercury in the solar wind [*see bottom illustration on page 45*]. A close passage in the wake of a planet provides the least ambiguous observational path for a flyby, particularly to test for the presence of anything resembling an earthlike magnetic field.

On the first flyby, when *Mariner 10* passed about 700 kilometers above the surface of Mercury, the appropriate instruments detected a weak magnetic field and a generally earthlike interaction with the solar wind. In order to get more definitive observations the spacecraft was targeted to make its third pass still closer to the surface (327 kilometers) on a track that carried it closer to the north pole. The third flyby confirmed the strength and orientation of the magnetic field that had been predicted by Norman F. Ness of NASA on the basis of the first flyby. Mercury evidently has a dipole magnetic field approximately aligned with the planet's spin axis. The strength of the field ranges from 350 to 700 gammas at the surface, or about 1 percent of the strength of the earth's surface magnetic field. Mercury's field is far stronger than the field found at either Venus or Mars, and a planetwide internal mechanism seems to be required for its generation. In addition Mercury is surrounded by a very thin envelope of helium gas, suggesting that the planet's magnetic field entraps helium nuclei from the solar wind and possibly from surface emanations as well.

The existence of a seemingly earthlike magnetic field on Mercury certainly provides independent evidence for a chemically differentiated planet with an iron core. Pictures acquired on the second

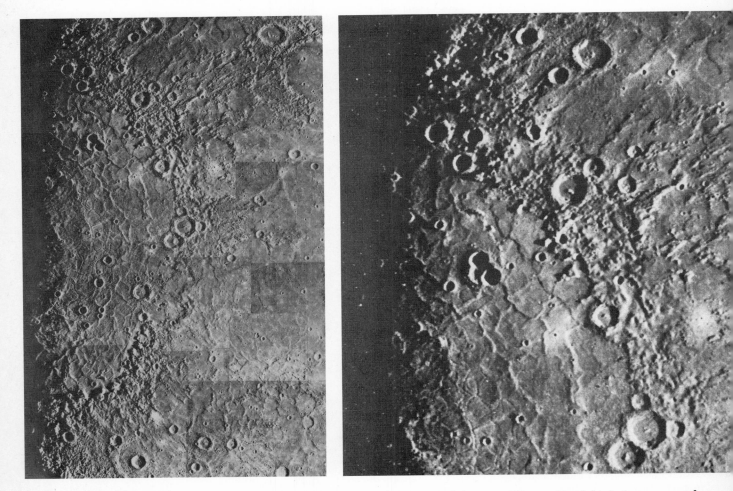

**CRINKLED FLOOR OF THE CALORIS BASIN** is shown with increasing resolution in the sequence of five pictures across these two pages. The first picture is a mosaic of views taken during *Mariner 10*'s first encounter with Mercury, looking back from a distance of about 75,000 kilometers. The second and third pictures were taken earlier in the same pass, when the spacecraft was closer to the planet. The fourth and fifth pictures were taken 352 days later, when the spacecraft made its third close approach to Mercury after

flyby, last September, showed that the lobate scarps seen in the first series of pictures continue across the south-polar region. Pictures from the first and second passes also show that small, steep-walled craters near the two poles have areas that are permanently shaded from the sun and thus constitute enduring "cold traps" for whatever volatile substances may have been released intermittently over Mercury's history. Those sites might be an exciting objective for close inspection in the 21st century.

The "outgoing" hemisphere of the planet photographed by *Mariner 10* as it "looked back" after its close approach exhibits a configuration of surface features totally different from those presented by the "incoming" hemisphere [*see illustration on page 38*]. The outgoing hemisphere shows large areas of relatively smooth plains that are clearly younger than most of the heavily cratered terrain visible in the incoming hemisphere. In addition there is a 1,400-kilometer basin left from a gigantic impact comparable to the one that gave rise to the Imbrium Basin on the moon. This prominent Mercurian feature, named the Caloris Basin because of its equatorial location near the "hot" 180-degree longitude of Mercury, is, like the Imbrium Basin, entirely filled with smooth plains material.

Other plains, less cratered and deformed, extend eastward and northward for thousands of kilometers. The *Mariner 10* television team has concluded that the temporal sequence of these plains, their variation in reflectivity and color, their size relations and their geographic association all point to a widespread episode of volcanism that followed the end of the period of heavy bombardment. The resolution provided by the pictures does not, however, reveal the surface morphology clearly enough to unambiguously identify the origin of the plains.

The situation is reminiscent of the debate about the origin of the smooth plains on the moon before the Apollo missions. It appears that some of the lunar plains were created by gigantic impacts rather than by volcanism. Hence the possibility remains that the material covering the Mercurian plains consists of massive sheets of ejecta from huge impacts, conceivably located on the hemisphere of Mercury that was in shadow during the three *Mariner 10* encounters. Whatever the origin of the plains, the fact that they have not been crumpled by internal activity and have not been modified by atmospheric erosion or deposition testifies to the remarkable quiescence of Mercury since the plains were made. Indeed, the overall similarity in the sequence of events that shaped the surface of Mercury and the moon is extraordinary and, to me, surprising, considering how very different the internal constitution of the two bodies must be.

A preliminary surface history of Mercury can be divided into five major sequences of events, each generally similar to those that shaped the moon. First,

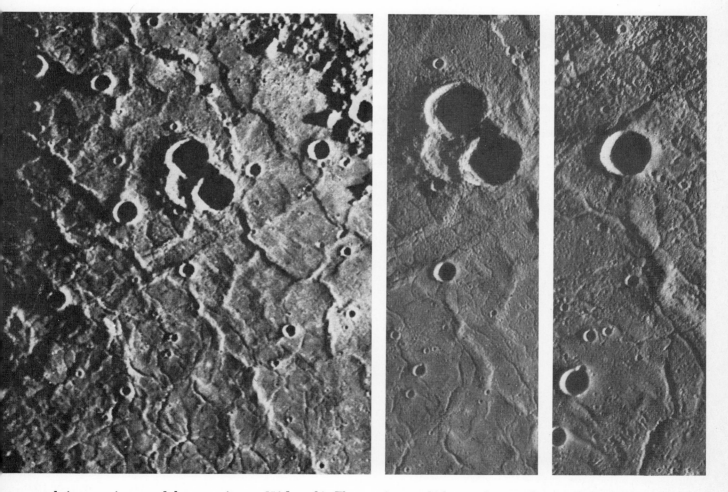

completing two trips around the sun on its own 176-day orbit. The fractures that are visible in the floor of the Caloris Basin vary in width from about eight kilometers down to about 450 meters, the limit of photographic resolution. The largest crater in the fifth picture, which was taken at a distance of 20,000 kilometers above the surface of the planet, is about 12.5 kilometers in diameter. Each of the two young craters in the prominent cluster of three craters that is seen in the other pictures is about 35 kilometers in diameter.

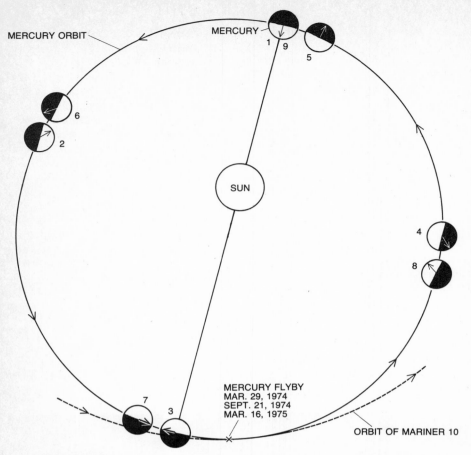

**PHENOMENON OF SPIN-ORBIT COUPLING** is what has locked Mercury's rotation period and orbital period in the ratio of two to three. In this diagram of Mercury's orbit the fixed arrow points toward one of the planet's two hot subsolar points, that is, the points on the equator that lie directly under the sun at alternate perihelions. The numbers give the sequential position of the planet in its orbit during two of its revolutions around the sun. *Mariner 10*'s encounters with Mercury take place at the point that is marked with an X.

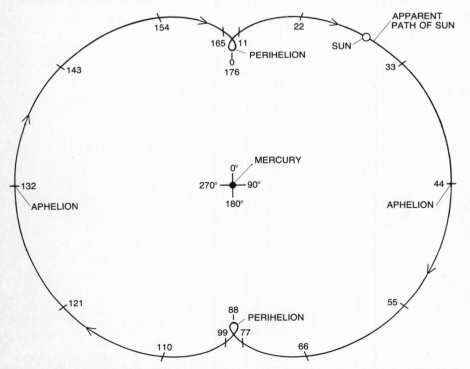

**SUN AS SEEN FROM MERCURY** appears to execute a loop at perihelion. Apparent position of the sun in relation to subsolar longitudes on the planet is marked off in 11-day intervals for two Mercurian years. Pattern of motion was described by S. Soter and J. Ulrichs.

the absence of recognizable volcanic, tectonic or atmospheric modification of the large old craters on Mercury implies that the mass of the planet was chemically fractionated into a large iron core enclosed by a thin silicate mantle well before any of the oldest craters were formed. Any atmosphere on Mercury must have escaped quite early or was never formed. The heat required for the chemical separation of iron and silicate phases on a global scale must also have dissipated early enough for the outer layers to become sufficiently rigid to maintain the topographic relief of the large old impact craters up to the present time.

In the second major epoch, following the initial period of accumulation and chemical segregation, Mercury must have gone through at least one period of crater obliteration, possibly through early volcanism, if we are to account for the absence of topographic scars of the accretion process. The smoothed surface, surviving in the plains between craters, records not only the terminal phase of heavy bombardment but also the global shrinking of the crust represented by the lobate scarps.

The third and most sharply delineated of the five major subdivisions of Mercury's surface history came toward the end of the heavy bombardment with the large Caloris Basin impact, seemingly the counterpart of the lunar impact that created the Imbrium Basin. The Mercurian collision gave rise both to the Caloris Basin and to the mountainous terrain surrounding it and extended areas of ejecta and sculpturing of the older surface. In addition, approximately antipodal to the Caloris Basin there is a peculiar lineated and hilly terrain, which Gault and Peter Schultz speculate may have been thrown up by the focusing of seismic energy from the Caloris Basin impact on the opposite side of the planet.

During the fourth phase, some time after the Caloris Basin impact, broad plains were created, probably as a result of widespread volcanism similar to the activity that gave rise to the lunar maria, or "seas." Bruce W. Hapke of the University of Pittsburgh argues that the colormetric and photometric evidence from *Mariner 10*, combined with previous telescopic results, suggest that the surface rocks on Mercury may be somewhat less rich in iron and certainly less rich in titanium than the rocks found on the maria of the moon, thus explaining the absence on Mercury of the sharp contrast in brightness levels between the lunar maria and the lunar highlands.

During the fifth phase of the surface history of Mercury, extending to the present time, little has happened other than a light peppering of impacts, many of which show conspicuous rays. The distribution of impact craters on the Mercurian plains is remarkably similar to the distribution both on the maria of the moon and on some of the oldest smooth plains of Mars. The similarity of impact-cratering rates on Mercury, the moon and Mars over the past three billion years or so came as a surprise to many, considering the great differences in the three bodies' distance from the sun and therefore the differences in the probability of their encountering interplanetary debris from the asteroid belt (which has long been thought to be the principal source of the impacting objects).

Thus *Mariner 10* has clearly established that the exterior of Mercury resembles the moon not just in topography but more surprisingly in surface history. And yet the interior constitution of Mercury appears to be more earthlike than that of any of the other planets. The paradoxical circumstance of Mercury's moonlike exterior and earthlike interior raises important questions not only about Mercury but also about the history and nature of the entire inner part of the solar system. Were the bombarding objects whose impacts are recorded on the surface of Mercury from the same family of objects that bombarded the moon as recently as four billion years ago? Or did each of the inner planets, including the moon, pass through separate periods of late bombardment that overlapped only slightly, if they overlapped at all, and in each case ceased abruptly and independently?

The *Mariner 10* pictures suggest to me and to Newell J. Trask, Jr., of the U.S. Geological Survey that the last great bombardment of Mercury, culminating with the Caloris Basin impact, may not have been part of a steadily decreasing flux of interplanetary debris but could have resulted from a discrete terminal episode of bombardment. George W. Wetherill of the University of California at Los Angeles considers it plausible that the bombarding objects involved in such an early discrete episode could have originated with a single object perturbed to pass near the earth or Venus from an initial orbit beyond Jupiter. Tidal disruptions on the earth or Venus might then conceivably have created a shower of bombarding objects that would have been rapidly swept up through collisions with the four inner

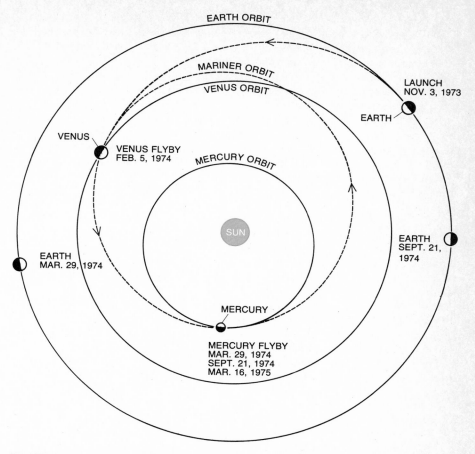

**TRAJECTORY OF MARINER** 10 was chosen so that the spacecraft was deflected toward Mercury by a precisely timed encounter with Venus three months after leaving the earth. It is as a result of this deflection that the orbit of *Mariner 10* takes it to the vicinity of Mercury every 176 days, or once every two Mercurian years. It was this happy choice of orbital characteristics that enabled *Mariner 10* to achieve three operational encounters with its target.

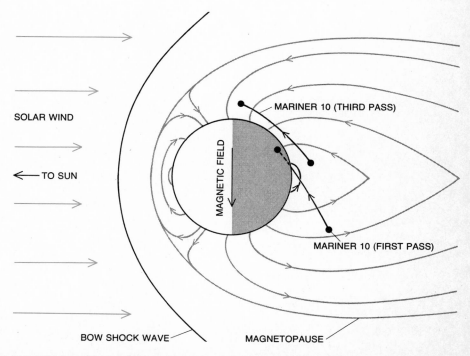

**MAGNETIC FIELD OF MERCURY**, detected in *Mariner 10*'s first encounter, was fully confirmed in the third encounter, when the spacecraft was targeted to pass closer to the north pole of the planet. The black dots indicate in planar projection when *Mariner 10* entered and left the planet's magnetic field on each pass. Field lines are distorted by pressure of charged particles in the solar wind. The polarity of Mercury's field is oriented like that of the earth.

planets. Such a concept would be consistent with recent controversial proposals that the moon was subjected to a similar terminal episodic bombardment about four billion years ago.

On the other hand, if the observed topography was created by a continuously declining bombardment of Mercury, the evidence may have been abruptly and effectively erased by an episode of enhanced crater obliteration. Hence the appearance of an episodic bombardment of the planet may be an illusion. The debate now developing over the early history of the inner solar system is reminiscent of an earlier debate between uniformitarians and catastrophists over the causes of the earth's geological features. There the uniformitarians won.

That Mercury has a dipole magnetic field aligned with its spin axis, very similar to the earth's field although weaker, is to me particularly unexpected. Granted that Mercury probably has a large iron core, the rotation of the planet is nevertheless so slow at present that before *Mariner 10*'s encounter with the planet no one thought that a Mercurian field might be generated by a fluid-dyna-

mo mechanism of the type postulated for the earth (in which the field arises from electric currents associated with fluid motions in the core of the spinning earth). And what about Venus, which presumably has a larger and hotter core than Mercury's and yet does not exhibit a significant planetary magnetic field? Furthermore, if there are fluid motions within the Mercurian core capable of generating the observed magnetic field, the core motions or the associated heat flow have not led to any recognizable deformations of the planet's surface layers.

A quite different explanation for the planet's field is that it is a fossil field of some kind that has persisted from a remote epoch. It seems unlikely, however, that in the billions of years since the hypothetical fossil field was created the temperature within the appropriate portion of Mercury's interior never rose above the Curie point (the temperature at which a substance loses its magnetism). Still a third possibility is that the field has somehow been induced as a result of Mercury's continued interaction with the solar wind. Preliminary examination of this concept suggests that

such a field would not exhibit symmetry around the rotation axis.

Perhaps the Mercurian magnetic field arises from causes still unimagined. Or perhaps we shall have to gain a deeper understanding of the mechanism of the earth's own field in order to explain how that mechanism could be reduced to the Mercurian scale. Whatever the outcome, it is fortunate that there is another magnetic field among the inner planets with which to compare and contrast the field of the earth. Further study of Mercury's field, as well as photographic mapping of the still unobserved hemisphere of the planet, would be a major objective of any orbiting satellite of Mercury.

*Mariner 10*'s mission to Mercury has completed the reconnaissance of the inner solar system and has further demonstrated that planetary exploration is full of surprises. The findings of *Mariner 10*, combined with the testimony of the lunar samples brought back by the Apollo astronauts, have made possible an enormous extension of knowledge and inference about Mercury. Out of the new observations is emerging a richer and more unified picture of the origin and evolution of all the planets, including our own.

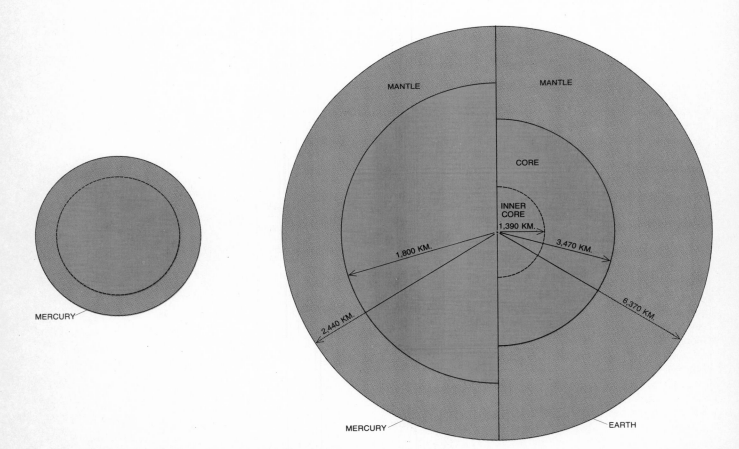

**CUTAWAY VIEWS OF MERCURY AND THE EARTH**, which are scaled to have the same outside diameter, show how much larger Mercury's iron core is thought to be compared with earth's core. Both of the cores are shown in dark color. Mercury's actual size in relation to the earth is shown by the small disk at the left. Mercury's iron core evidently contains 80 percent of the planet's mass. Therefore the iron core must have a radius of at least 1,800 kilometers, which would make core alone slightly larger than the earth's moon.

# 5

# VENUS

# Venus

ANDREW and LOUISE YOUNG

*It is cratered like the rest of the inner planets, but its surface has been transformed by its dense and cloudy atmosphere. The clouds trap sunlight to maintain a temperature of 900 degrees Fahrenheit*

Until about 1960 Venus was commonly regarded as "the earth's twin." It is our nearest neighbor in the solar system, and it is approximately the same size and density as the earth. Because Venus is closer to the sun it intercepts twice as much sunlight as the earth, but a thick, global blanket of clouds reflects most of this light, so that the planet absorbs only about as much solar energy as the earth. On the basis of these similarities many observers concluded that conditions on the surface of Venus must resemble those on the earth. For example, it was thought that the length of Venus' day was about the same as the earth's, and that its atmosphere was made up of the same gases. The clouds were naturally supposed to consist of water, as the earth's do.

The work of planetary scientists over the past 40 years has shown that this conception of Venus was thoroughly mistaken. Although much remains to be learned about the planet, we now know that it is not at all like the earth. Venus is hot: at the surface the temperature reaches 750 degrees Kelvin (about 900 degrees Fahrenheit). The atmosphere is made up not of nitrogen and oxygen but of carbon dioxide, and the atmospheric pressure at the surface is 90 times that on the earth. There is little water, but the clouds contain corrosive compounds such as sulfuric acid. The rotation of Venus is very slow, one turn requiring about eight earth months, and it is in a retrograde direction, so that if the sun were visible through the clouds, it would rise in the west and set in the east.

Early misconceptions about Venus are readily pardoned; although the planet comes closer to the earth than any other, it is one of the most difficult to study. As it is seen from the earth, it is never far from the sun; as a result it must be observed either in the daytime, when the sky is bright, or in the early evening or morning, when the planet is near the horizon and we must look through a great thickness of our own atmosphere. Moreover, the surface of Venus is at all times obscured by its cloudy atmosphere. At visible wavelengths the clouds themselves appear to be featureless.

Since visual observation seems so unpromising, recent research has concentrated on other methods of studying the planet. Radio waves penetrate the Venusian clouds, and large features of the planet's surface have now been mapped by radar. Photographs made at ultraviolet wavelengths have revealed structures in the clouds that cannot be seen with visible light. From the many wavelengths in reflected sunlight we can derive not only a photographic image but also information on the composition of the atmosphere (from the spectrum of absorption lines), on the motion of the atmosphere (from shifts in the wavelength of the absorption lines) and on the properties of particles in the clouds (from the polarization of the light). Finally, eight spacecraft have passed close to the planet or descended into its atmosphere, providing direct measurements and, most recently, photographs at higher resolution than can be attained from the earth. Two more planetary probes are on their way to Venus now.

In 1897 Edward Emerson Barnard of the Yerkes Observatory wrote: "No other object has caused more controversy and produced more varied testimony in the determination of its rotation period than the planet Venus. This rotation controversy has raged for upwards of two centuries." A few years later the first attempts were made to deduce the rotational speed from measurements of the Doppler effect on light reflected from Venus. The method depends on detecting the change in the wavelength of spectral absorption lines; a comparison of the shift in wavelength at the center of the planet's disk and at the limb gives the rotational speed. The change in wavelength is small (it is proportional to the ratio of the rotational speed to the speed of light) and measuring it accurately is a delicate task. By 1956 Robert S. Richardson of the Mount Wilson and Palomar Observatories had obtained a value suggesting that Venus rotates slowly in the retrograde direction.

The controversy did not end with

SURFACE FEATURES OF VENUS, including a large crater, are visible in the radar-reflectivity map on the opposite page. The map, which covers a small area just north of Venus' equator, is a detail of a somewhat larger map made by Richard M. Goldstein and his colleagues at the Jet Propulsion Laboratory of the California Institute of Technology. It was made by transmitting a beam of microwaves to the planet and detecting the echo with two radio telescopes at Goldstone, Calif., operated as an interferometer. The pattern of dots results from the digital method of extracting information from the reflected signal. Apparent shadows on the surface are artifacts of the technique; the brightness of an area represents not its illumination but the amount of radio-frequency energy it reflects. Many of the surface features are large in extent, but they have a small range of heights and depths. The large circular form is a crater whose diameter is about 160 kilometers but whose rim is only about 500 meters high. The interior of the crater does not appear to be much below the level of the surrounding terrain. The surface of Venus cannot be observed by optical methods because a thick blanket of clouds covers the entire planet. The corrosive substances thought to make up the clouds may be responsible for the shallowness of surface features.

Richardson, however. In the early 1960's Bernard Guinot of the Haute-Provence Observatory in France was measuring the orbital speeds of the inner planets and found that his measurements for Venus were inconsistent: the velocity varied when it was measured at different places on the planet's disk. The inconsistency could be resolved by assuming that the planet was rotating in a retrograde direction, but only if the rotation was much faster than Richardson's data indicated: about 100 meters per second at the equator. That speed corresponds to one rotation approximately every four days.

Guinot's discovery was supported by observations made by another method. Examining ultraviolet photographs of Venus, the French astronomer C. Boyer found that dark features visible in the clouds sometimes reappeared after four days. The first extensive studies of Venus at infrared and ultraviolet wavelengths had been made by Frank E. Ross at the Mount Wilson Observatory in 1928. The infrared pictures were featureless but the ultraviolet ones showed broad, hazy markings of low contrast. Ross had looked for periodic changes in these

CLOUDS IN THE ATMOSPHERE of Venus reveal a complex pattern of bright and dark swirls. The image is a mosaic of 56 photographs made by one of two television cameras aboard the U.S. spacecraft *Mariner 10,* which passed within 5,800 kilometers of Venus on February 5, 1974. The photographs were made at ultraviolet wavelengths; in visible light the cloud tops are a featureless pale yellow. The patterns evident at ultraviolet wavelengths are presumably a product of the uneven distribution of a substance that absorbs ultraviolet radiation. The image has been processed to enhance contrast and to emphasize smaller features. The mosaic was only recently completed by workers at the Jet Propulsion Laboratory. It has a resolution of about seven kilometers, substantially better than that of earlier mosaics, which were constructed from fewer photographs made at greater distances from the planet.

cloud markings but had found none. Boyer did see changes suggestive of movement, and he was stimulated by Guinot's work to make more photographs. He found that an apparent retrograde motion could be detected in exposures made only a few hours apart. The observed motion was consistent with a rotational velocity of 100 meters per second. Similar cloud motions have been observed in more detailed photographs made by the spacecraft *Mariner 10* [*see illustration at right*].

Spectroscopic and photographic techniques measure the rotational velocity only of Venus' atmosphere, and indeed only of the upper layers of the atmosphere. Through radar measurements, which also depend on the Doppler effect, it is possible to detect the rotation of the solid body of the planet. A radar determination of Venus' rotational speed was published in 1962. It confirmed that the planet's spin is retrograde but indicated that the rotation is very slow. Venus makes one complete rotation in 243 earth days.

This rotational period suggests a relation between the motions of the earth and Venus that may not be coincidental. At every inferior conjunction, that is, whenever Venus lies between the earth and the sun, the planet presents the same side to the earth. It is possible that the rotation of Venus has been "captured" by tidal forces generated when the two planets approach closely. That is how the moon has come to rotate with the same side constantly facing the earth. Even at inferior conjunction, however, the earth and Venus are much farther apart than the earth and the moon, and the tidal forces they exert on each other are therefore much smaller. If Venus' rotation has been entrained in this way, its mass must be distributed asymmetrically; at the equator the planet must deviate from the form of a perfect sphere by at least one part in 10,000. The careful tracking of spacecraft flying past Venus at close range would have revealed any irregularities in the planet's gravitational field much exceeding that value, but none have been detected. The issue will probably not be decided until a spacecraft has been put in orbit around Venus.

The radar measurements of Venus' solid-body rotation are quite accurate. On the other hand, the much higher velocities implied by spectroscopic observations and by the apparent movement of cloud features across the planet's disk cannot be ignored. If the two kinds of measurement are both correct,

we must explain how a slowly rotating planet can have a rapidly rotating atmosphere; in other words, we must account for enormous winds.

Recent spectroscopic observations by Wesley A. Traub and Nathaniel P. Carleton of the Center for Astrophysics in Cambridge, Mass., suggest that winds on Venus vary from zero to more than 100 meters per second. Data from the Russian Venera spacecraft that entered Venus' atmosphere have been interpreted as indicating wind speeds of from 10 to 100 meters per second. In the earth's atmosphere such high winds are encountered only in narrow jet streams. Jet streams could not, however, account for the rapid atmospheric movements observed on Venus, since the Venusian winds seem to involve large regions of the planet.

The apparent agreement between the spectroscopic and the photographic evidence has recently been questioned. The best spectroscopic measurements seem to imply a somewhat slower atmospheric circulation, corresponding to a rotation period of about six earth days. That motion is still far more rapid than the 243-day period of the planet itself, but it differs significantly from the four-day rotation period usually observed in the cloud movements. It is not certain that the two techniques measure the same phenomenon.

Theoretical attempts to explain the generation of the winds have produced several possible mechanisms, such as convection caused by the uneven heating of the day and night sides of the planet. None of them, however, has been shown to be capable of explaining velocities greater than a few meters per second. Furthermore, any comprehensive theory of atmospheric circulation on Venus must account for other, more complex phenomena. For example, infrared observations have revealed that the cloud tops move vertically as well as horizontally, and that the vertical movement is roughly periodic. The period is variable, but it usually lies between four and six earth days. It is therefore possible that the shifting features do not represent bulk movements of gases at all but are waves propagating through the atmosphere.

Because the surface of the planet is so difficult to observe, the study of Venus has been largely the study of its atmosphere. Of particular interest is the chemical composition of the atmosphere and of the clouds. Traditionally the most important method of investigating the chemistry of the atmosphere has been spectroscopy. In the sunlight reflected

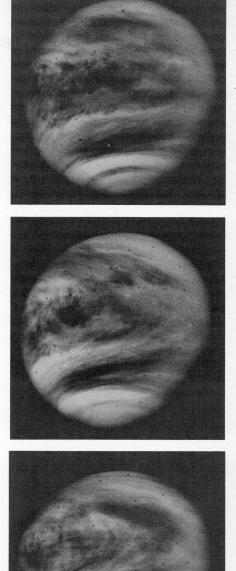

APPARENT MOTION of clouds high in the atmosphere of Venus can be detected in a series of ultraviolet photographs made by *Mariner 10*. The photographs were made at intervals of seven hours on February 7, 1974, two days after the spacecraft's closest approach to the planet. A prominent dark marking is near the center of the disk in the first exposure; it moves rapidly to the left and in the third photograph is nearly at the limb. Similar motions have been observed from the earth, and suggest that the upper atmosphere may rotate in a retrograde direction with a period of about four days. Since the solid body of the planet is known to rotate much more slowly, the observed atmospheric motions are interpreted as high-speed winds extending over large areas.

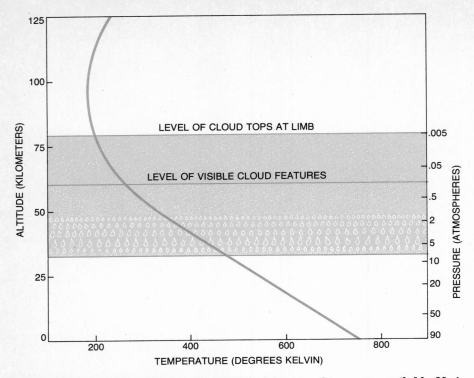

**TEMPERATURE AND PRESSURE PROFILES** of the atmosphere were compiled by Mariner and Venera spacecraft. They indicate that below an altitude of about 90 kilometers temperature steadily increases. The pressure reaches one atmosphere (the mean pressure at sea level on the earth) near 50 kilometers. At the surface the temperature is about 750 degrees Kelvin and the pressure is about 90 atmospheres. The temperature gradient (the change in temperature with altitude) is limited by convection to about 10 degrees per kilometer.

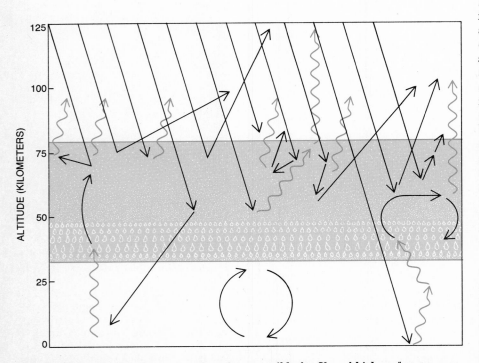

**RADIATION** trapped by the atmosphere is responsible for Venus' high surface temperature. Much of the incident solar radiation (*straight arrows*) is reflected by the clouds. About 20 percent is absorbed at the cloud tops, mainly in the ultraviolet (by an unidentified absorber) and in the near infrared (by carbon dioxide). The heat absorbed by the surface is reradiated in the thermal infrared (*wavy arrows*), a spectral region of somewhat longer wavelengths than the near infrared. Radiation in the thermal infrared is absorbed by gases below the clouds and by the clouds themselves, and hence cannot directly escape into space. Within the atmosphere heat is transported principally by convection (*curved arrows*).

by the planet the narrow absorption lines characteristic of particular molecules can be detected and the molecules identified.

The first substance detected on Venus was carbon dioxide, discovered by Walter S. Adams and Theodore Dunham, Jr., of the Mount Wilson Observatory in 1932. Within two years Arthur Adel and V. M. Slipher of the Lowell Observatory had shown that at the earth's atmospheric pressure a column of carbon dioxide at least two miles long was needed to produce absorption lines as intense as those in the Venus spectrum. That is in reasonable agreement with modern estimates of the thickness of the layer of absorbing gas above the clouds. Today Venus is the best available source of carbon dioxide spectra [*see illustration on page 54*]. In spectra made at high resolution, such as those obtained by Pierre and Janine Connes in France, more than 5,000 absorption lines have been identified. Most of them have never been observed elsewhere; they are produced by carbon dioxide molecules containing rare isotopes such as carbon 13, oxygen 17 and oxygen 18.

In the early 1960's William M. Sinton of the Lowell Observatory and V. I. Moroz of the Shternberg Astronomical Institute's South Station in the Crimea independently discovered evidence for a second molecule in infrared spectra of Venus: carbon monoxide. They also showed that Venus reflects very little sunlight at wavelengths between three and four microns, which lie in the near infrared. Moroz pointed out that the substance responsible for this absorption band could not be water, but that it must serve to trap heat in the lower atmosphere of Venus, a role that in our own atmosphere is played by water. Sinton and John Strong of Johns Hopkins University also found a broad absorption feature at a wavelength of 11.2 microns, in the thermal infrared, which remained unexplained for more than a decade.

As recently as 10 years ago carbon dioxide and carbon monoxide were the only substances whose presence on Venus had been established. Evidence for water vapor had long been sought, and there had been numerous reports of its discovery, all of them now thought to have been spurious. The first authentic observation of water vapor on Venus was probably made by Ronald A. Schorn and his colleagues at the McDonald Observatory in the late 1960's. Since then the abundance of water vapor in Venus' upper atmosphere has been measured frequently by Edwin S. Barker of the Mc-

Donald Observatory. The relative humidity rarely reaches 1 percent.

Absorption spectra of Venus have also been scrutinized for evidence of a major constituent of the earth's atmosphere, oxygen. It has not been found, and its absence presents an enigma. The carbon monoxide on Venus is presumably formed in the upper atmosphere when carbon dioxide is dissociated by ultraviolet light. Oxygen is an inevitable by-product of this reaction, and since molecular oxygen is diatomic there should be half as many $O_2$ molecules as there are CO molecules. That is the ratio on Mars, another planet with an atmosphere composed mainly of carbon dioxide. On Venus oxygen is at least 50 times less abundant than carbon monoxide.

The oxygen deficiency might be explained if oxygen is quickly transported to the lower atmosphere, where it could combine with other substances, such as sulfur. If that is the case, the mixing of gases in the stratosphere would have to be much more efficient on Venus than it is on the earth.

Recently it has become possible to study Venus' atmosphere by methods that are more direct than spectroscopy. On two of the Venera spacecraft atmosphere-sampling instruments have been taken into the atmosphere itself. Unfortunately the instruments were capable of detecting only a few gases, and they could not report on any unexpected molecules. Moreover, the presence of unexpected substances in the atmosphere could interfere with the analysis and result in incorrect measurements.

In spite of these difficulties the Venera craft determined the amount of carbon dioxide in Venus' atmosphere. It was shown to be the major constituent, making up about 97 percent of the bulk of the atmosphere, rather than a minor one, as some theorists had argued. On the other hand, the Venera data heightened confusion over the amount of water vapor on Venus by indicating that there is almost 1,000 times as much as spectroscopic measurements indicate. Subsequent studies at radio wavelengths have established once again that there is no more than .1 or .2 percent water vapor in the lower atmosphere, and the true value is probably closer to .01 percent. The cloud tops are drier still. We now believe the Venera measurements may have been contaminated by material in the Venusian clouds.

Determining the composition of the clouds is more difficult than identifying the atmospheric gases because liquid and solid materials do not produce the narrow spectral lines that uniquely identify gases. When the clouds are observed in the visible part of the spectrum, they are a uniform pale yellow. Only at ultraviolet wavelengths are features visible; in ultraviolet photographs the entire planet appears gray, but some regions are darker than others. If the markings are caused by the uneven distribution of some material that absorbs ultraviolet radiation, we must ask what the material is and why it is distributed unevenly.

The yellow color of the clouds has long been cited as evidence that they do not consist of water, for the obvious reason that terrestrial water clouds are white. Ross early suggested that they might be clouds of yellow dust, like those occasionally seen on Mars. Since then a diverse array of materials has been proposed: plastics, salts, liquid mercury, compounds of mercury, hydrocarbons, polymers of carbon suboxide ($C_3O_2$), partially hydrated ferrous chloride, hydrochloric acid. The candidate that is most widely accepted today is sulfuric acid. It is as powerless as water to explain the yellow color of the clouds, but it does provide an economical explanation for many other observations, so that its lack of color is no longer considered a fatal flaw.

Among the more exotic materials proposed for the clouds only one has been detected spectroscopically. It is hydrogen chloride, and it was found along with hydrogen fluoride by William S. Benedict of the University of Maryland in the spectra recorded by the Conneses. Both gases are highly corrosive; when they are dissolved in water, they yield hydrochloric acid and hydrofluoric acid. Their abundance is too low for them to be the principal constituents of the clouds, but that they should be present in the atmosphere at all was a surprise.

Such strong acids could not survive for long in the earth's atmosphere; they would react with rocks and other materials and soon be neutralized. That they exist on Venus implies that in some crucial ways conditions there differ from those on the earth. One important difference is in surface temperature.

The temperature of Venus' upper atmosphere can be measured from the earth with reasonable accuracy. The

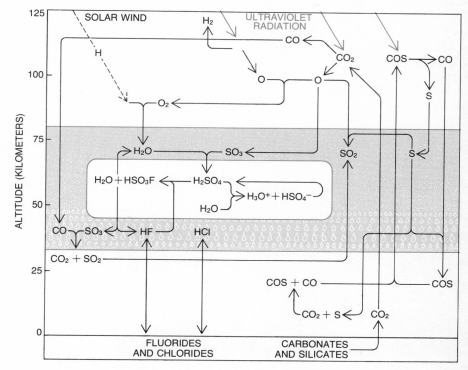

ATMOSPHERIC CHEMISTRY is strongly influenced by conditions at the surface. Carbon dioxide is liberated from rocks by high temperature, and above the clouds some of it is broken down into carbon monoxide (CO) and oxygen ($O_2$). The oxygen may enter a series of reactions that culminates in sulfuric acid ($H_2SO_4$), which, with traces of hydrochloric acid (HCl) and hydrofluoric acid (HF), is believed to make up the cloud particles. Elemental sulfur and compounds of sulfur, carbon and oxygen may also be present. The reactions above the clouds are generally photochemical; those below are thermochemical. Reactions that involve molecules in the liquid phase are indicated by a white box; all others are between gaseous molecules. Water is scarce; some might be lost through the escape of hydrogen into space and some might be formed from hydrogen captured from the solar wind.

thermal emission of the clouds at infrared wavelengths can be detected, and the temperature can also be deduced from the molecular absorption spectrum in reflected solar light. Both techniques indicate that the temperature at the cloud tops is about 250 degrees K., which is below the freezing point of water (273 degrees K.).

These values, of course, apply only to the upper atmosphere. We now know that the surface of the planet is much hotter. The temperature in regions below the clouds was first measured in the mid-1950's by detecting thermal emission at radio wavelengths. The initial studies indicated a surface temperature of about 600 degrees K., a value so high that most astronomers believed it must be erroneous. Perhaps because of the widespread notion that Venus should be earthlike, the high values were not universally accepted until *Mariner 5* and *Venera 4* recorded temperature profiles of the atmosphere in 1967. From direct measurements made by the later Venera probes the surface temperature is now known to be about 750 degrees K.

From the large difference in temperature between the surface and the cloud tops, some of the physical properties of the atmosphere can be deduced. In particular the atmosphere must be deep and therefore massive, because a shallow atmosphere could not sustain such a large temperature gradient. On the earth the maximum gradient is about 10 degrees K. per kilometer; if a larger gradient were to develop, it would quickly be eliminated by convection. The same convective process must also limit the temperature gradient on Venus.

The high temperature at the surface of Venus also has important effects on the chemistry of the atmosphere. As John S. Lewis of the Massachusetts Institute of Technology has shown, small amounts of hydrogen chloride and hydrogen fluoride could be "cooked out" of surface rocks at high temperature, in the same way that these acids are evolved in volcanic gases on the earth. A number of assumptions are implicit in this hypothesis: that the rates of chemical reactions at the surface are high, that the atmosphere and the surface are in chemical equilibrium and that the effects of circulation in the atmosphere are small enough to be neglected. High temperature can surely be expected to accelerate chemical reactions, but the actual rates have not yet been measured, and the other assumptions cannot be tested at all. In spite of these uncertainties knowledge of the thermal environment of Venus provides a reasonable "shopping list" of molecules to look for in the atmosphere.

The temperature and other physical properties of the atmosphere can also help to identify the materials that make up the clouds. One of the most powerful means of analyzing the cloud particles has proved to be the study of the polarization of sunlight reflected from the clouds. The polarization was first observed by the French astronomer Bernard Lyot in the 1920's; more recent measurements have been made by Audouin Dollfus in France and by Tom Gehrels and D. L. Coffeen, both of the University of Arizona. From theoretical calculations J. E. Hansen and his colleagues at the Goddard Institute for Space Studies in New York have shown that the polarization can be explained if the cloud particles are spherical, with a radius of about a micron, if they have a narrow range of sizes and a refractive index for green light of about 1.44 and if the refractive index varies with wavelength in the normal way.

That list of specifications imposes several constraints on the choice of cloud constituents; initially it eliminated all the candidates. The spherical shape of the particles implies that they are liquid droplets rather than solid crystals. Water is excluded because its refractive index (1.33) is too low, and also because it would freeze at the temperature of the cloud tops and therefore would not have a spherical form. There was an attempt to make hydrochloric acid droplets conform to the requirements of Hansen's calculations, but it proved impossible to make the refractive index high enough under conditions at the cloud tops. Most other inorganic substances were eliminated because their refractive index is too high.

Sulfuric acid was proposed independently and almost simultaneously by Godfrey Sill of the University of Arizona and by the authors and later by Ronald G. Prinn of M.I.T. Prinn realized that he could have predicted the presence of sulfuric acid in the clouds several years before, when he had discovered a chain of photochemical reactions that culminated in the formation of sulfur trioxide ($SO_3$), the anhydride of sulfuric acid. Because sulfur trioxide has not been observed on Venus, he had assumed it was destroyed as fast as it formed. Actually any sulfur trioxide formed in the cooler regions of the atmosphere would immediately combine with water vapor to form a sulfuric acid haze.

The sulfuric acid hypothesis explains many of the observations that have baffled astronomers studying Venus over the past 15 years. Water solutions containing more than 70 percent sulfuric acid have the proper refractive index, and they are liquid at 250 degrees K. The failure of spectroscopic methods to detect gaseous sulfur compounds (such as $SO_3$) in the atmosphere is readily explained, since cold sulfuric acid has a negligible vapor pressure of sulfur-bearing molecules. Finally, sulfuric acid can account for some of the perplexing fea-

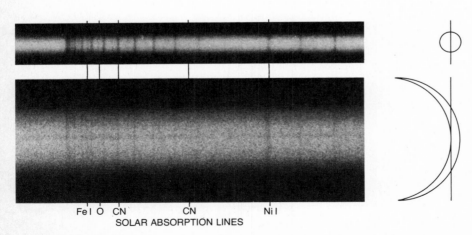

**ABSORPTION SPECTRA** of the atmosphere contain numerous lines associated with carbon dioxide. The spectra were recorded near superior conjunction (*top*) and near inferior conjunction (*bottom*). The diagrams at right indicate the phase and apparent size of the planet; the vertical lines indicate the position of the slit of the spectrograph on the planet's disk. A few of the absorption lines are part of the solar spectrum, but all the rest result from the absorption of solar light by molecules of carbon dioxide, which constitutes about 97 percent of Venus' atmosphere. These spectra cover only a narrow range of wavelengths in the near infrared. There are many additional carbon dioxide lines in other regions of the spectrum; indeed, Venus is the best available source of carbon dioxide spectra.

Fe I  O  CN          CN          Ni I
SOLAR ABSORPTION LINES

tures of Venus' absorption spectrum. Water solutions of the acid ionize almost completely to $H_3O^+$ and $HSO_4^-$, so that few water molecules would be left intact; the atmosphere would therefore appear to be very dry. Furthermore, the $H_3O^+$ ion absorbs strongly between three and four microns and the $HSO_4^-$ ion absorbs at 11.2 microns, thus explaining the anomalous spectral features discovered by Sinton, Moroz and Strong.

If allowance is made for other acids dissolved in the droplets, a concentration of about 80 percent (by weight) seems consistent with the data. Such a solution contains between one water molecule and two water molecules for each molecule of sulfuric acid. There are reasons to suppose that the number of droplets per unit volume gradually increases with depth, and that a few tens of kilometers below the cloud tops the droplets should approach one another closely enough to coagulate. Where the pressure reaches a few atmospheres a sulfuric acid rain may be falling. Both *Mariner 5* and *Mariner 10* detected an absorber of microwave radiation at an altitude below 50 kilometers, and earth-based microwave observations by Michael A. Janssen of the Jet Propulsion Laboratory of the California Institute of Technology indicate that the absorber consists of particles rather than molecules in the gaseous state. Solutions of sulfuric acid have a high electrical conductivity and could produce these effects if enough large drops are present.

As a drop of sulfuric acid falls through hotter air, water evaporates from its surface and the acid becomes more concentrated. Because of the great affinity of the acid for water, concentrated solutions boil only at high temperature, higher than the boiling point of the pure acid. The solution with the highest boiling point contains more than 98 percent acid and has equal vapor pressures of acid and water. *Venera 8* apparently detected the base of the clouds at an altitude of about 35 kilometers, where the temperature is about 400 degrees K. If that altitude is correct, then at the base of the clouds water vapor and sulfuric acid vapor must each have an abundance of about 100 parts per million.

A rain of hot, concentrated sulfuric acid is by itself an intimidating prospect for the designer of a spacecraft, but there are worse possibilities. The small amount of hydrogen fluoride in Venus' atmosphere could react with the sulfuric acid to yield fluorosulfuric acid ($HSO_3F$). Fluorosulfuric acid is the strongest of the simple mineral acids; it attacks most

common materials and dissolves sulfur, mercury, lead, tin and most rocks. The rain on Venus may be the most corrosive fluid in the solar system.

Although sulfuric acid successfully accounts for a great many of the properties of the Venusian clouds, there is still at least one important observation it cannot explain: the yellow color of the planet. The color must be produced by some substance that absorbs the shorter wavelengths, mainly the blue and the ultraviolet. Sulfuric acid does not qualify, but no other substance has gained general acceptance. The problem is complicated by the variety of compounds formed by hydrogen, oxygen, sulfur, chlorine and fluorine, which are all known to be present. Perhaps the absorber of the blue and ultraviolet wavelengths is some compound of these elements or perhaps it is elemental sulfur.

Whatever the composition of the clouds their influence on the climate and weather of Venus is fundamental. The clouds block both incoming solar radiation and outgoing thermal radiation. The incoming sunlight is mainly scattered, however, whereas the outgoing heat—at infrared wavelengths—is strongly absorbed by sulfuric acid. The clouds are thus much more effective in preventing the escape of heat than they are in blocking its entry. They contribute to the "greenhouse effect" that is thought to be responsible for the planet's high temperature.

Measurements made by *Venera 8* showed that about 1 percent of the sunlight incident on Venus reaches the surface. It is about as dark there as it is on the earth on a dark, rainy day. Below the clouds the carbon dioxide atmosphere is nearly opaque to infrared wavelengths, so that the absorbed heat is trapped at the surface. Since the heat cannot escape by direct radiation, it must be transported by convection to the cloud tops, where it can be radiated into space. Temperature profiles recorded by entry probes show a convective temperature gradient from the surface to the cloud tops.

At the cloud tops about 20 percent of the incident sunlight is absorbed, half in the near-infrared carbon dioxide bands and half by the unidentified ultraviolet absorber. Because most of the energy input to the planet is received at that altitude, it is there that weather systems should be most active. Moreover, because the ultraviolet absorber is one of the principal means of heating the atmosphere, the four-day variations in ultraviolet cloud features might be re-

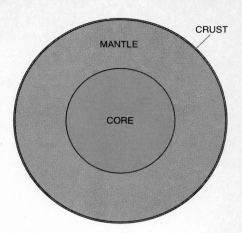

INTERIOR OF VENUS is supposed to be much like that of the earth. If the two planets formed in the same region of the developing solar system, their elemental compositions should be similar, and since their size and mass are nearly the same they are believed to have similar structures. Venus is therefore thought to have a liquid core, a mantle and a rocky crust. The size of the core and the thickness of the mantle and the crust are not known, although there is reason to suspect that Venus' core is substantially smaller than the earth's. The hypothesis that Venus' interior resembles the earth's is supported by a single measurement, made by *Venera 8*, indicating that the radioactivity of rocks in Venus' crust is similar to that of terrestrial granites. On the other hand, Venus has much less water than the earth, and it apparently has no magnetic field. The absence of a magnetic field might be explained by the slowness of Venus' rotation.

flected in the pattern of the planet's weather.

We have monitored the amount of gas above the clouds and measured its temperature, and have attempted to correlate changes in these parameters with the apparent period of rotation of the ultraviolet features. We have found that the amount of absorbing gas does vary in a four-day cycle, but on any given day it is nearly uniform over the entire planet; it shows no relation to the bright and dark cloud features seen in ultraviolet photographs. Moreover, the changing temperatures seem to be related neither to the height of the cloud tops nor to the movement of the ultraviolet features. The lack of any correlation is puzzling: the ultraviolet features seem to be painted on an otherwise featureless globe.

The surface and the interior of Venus remain hidden from us, but we are not altogether ignorant of their nature. Radar mapping has provided a preliminary picture of the planet's topography, and a few observations made from spacecraft have helped to establish a foundation for theories of its internal structure.

The radar mapping of the planets cannot be accomplished by the familiar scanning technique employed in terrestrial radars. Because the angular size of the planets is so small, the radar beam from a single antenna covers the entire disk of the planet, and the position on the surface from which a particular part of the returning signal was reflected must be deduced from a precise analysis of the delay and Doppler shift of the echo. Signals returning from the edge of the disk arrive slightly later than those from the center, and the extent to which frequencies are shifted by the rotational motion of the planet varies with both latitude and longitude. By connecting two or more antennas so that they can operate as an interferometer part of the planet's disk can be resolved.

Many areas on the surface have already been characterized by this method as being rough or smooth and high or low. On the most recent maps a few craterlike forms are visible [see illustration on page 48]. For the most part, however, Venus appears to be much smoother than the moon or Mars or Mercury. It is also remarkably flat, with few features more than a mile high. The scarcity of small craters can be attributed to Venus' dense atmosphere; meteoroids smaller than a few hundred feet in diameter could not penetrate it. The flatness of the terrain may also be a consequence of atmospheric phenomena. In the hot, dense and corrosive atmosphere weathering may be rapid and surface features may be quickly eradicated. This hypothesis requires, however, that the surface winds be strong enough to erode the terrain. In spite of the gales aloft, winds at the surface of Venus seem to be very gentle, and they may not be able to pick up and transport dust.

Most theories of the internal structure of Venus begin with the assumption that the planet is fundamentally similar to the earth, that is, it has a liquid core, a mantle and a rocky crust. There is a cosmological argument favoring this assumption: If the two planets condensed in the same region of the primitive solar nebula, they ought to have roughly the same elemental composition. The fact that they have approximately the same size and density lends support to the argument.

One possible objection to this hypothesis is Venus' lack of water. Another is the apparent absence of a magnetic field. Measurements made by *Venera 4* indicate that if a field exists, it is weaker than the earth's by a factor of at least 10,000. The absence of a magnetic field might be explained, however, without revising the theory of planetary structure. The earth's field is thought to be produced through the influence of the planet's rotation on motions within the liquid core. If that is the case, Venus may not have a field simply because it rotates very slowly.

Much more information about the solid body of the planet could be revealed by a direct examination of the rocks in its crust. If Venus formed in the same way as the earth, and if its atmosphere was produced by the escape of gases from its interior, the rocks should be strongly differentiated; if its evolution has not proceeded as far as the earth's, the rocks might be more like lunar basalts. As yet only one relevant observation has been made: *Venera 8* measured the radioactivity of the Venusian surface and indicated that it is comparable to that of terrestrial granites. This result suggests that the evolution of the crust has been earthlike, but it would not be prudent to draw a conclusion from a single measurement. Two Russian spacecraft that were launched in June and are expected to land on Venus in October may provide additional data.

Any speculation on the evolutionary history of Venus must immediately confront the problem of its atmosphere: Why should it be 90 times as massive as the earth's? We have a tentative answer to the question, but it leads to other and perhaps deeper enigmas.

There is about as much carbon dioxide on the earth as there is on Venus, but under terrestrial conditions most of it is bound as carbonates in rocks. It is only because the surface temperature of Venus is so high that the gas there remains in the atmosphere. This is hardly a satisfactory explanation, however, since it is the massive atmosphere that maintains the high temperature.

The problem is therefore one of how Venus acquired its atmosphere, not how it maintains it. A proposal that is widely accepted today depends on the greenhouse effect. It assumes that Venus was once more earthlike, that is, cooler and with a thinner atmosphere. Solar heat trapped by the atmosphere then raised the surface temperature. Eventually, because Venus is closer to the sun than the earth is, the boiling point of water was exceeded; the substantial quantity of water vapor thus added to the atmosphere greatly increased its opacity to infrared radiation and thereby produced a large increase in temperature. Finally the surface became hot enough to drive the carbon dioxide out of carbonates in rocks.

A few billion years from now the sun will grow brighter and the earth may acquire the massive atmosphere and high surface temperatures that Venus has today. At least in the initial stages of this process, however, when the oceans boil, the earth's atmosphere will be very different from what Venus' is now. In particular it will contain vast quantities of water, since the mass of water in the earth's oceans is much greater than the amount of carbon dioxide in its rocks. Venus, on the other hand, is quite dry.

If one assumes that Venus once had as much water as the earth has now, it is necessary to explain how all but one part per million of it was lost. There is a known mechanism by which a planet with abundant water could lose a large portion of it: water vapor in the upper atmosphere could be dissociated by ultraviolet radiation and the hydrogen could be lost to space, either by thermal escape or through the influence of the solar wind. That effect, however, could not produce an atmosphere so thoroughly desiccated as Venus' is. Of the water Venus has today, very little reaches the upper atmosphere and therefore it is not dissociated; at the present rate Venus would not have lost a significant amount in the history of the solar system.

Alternatively, one could assume that in the beginning Venus was entirely dry. The small quantity of water found there today could have been formed from hydrogen in the solar wind, which strikes the atmosphere directly because the planet has no magnetic field. That assumption, however, merely solves one problem by creating another, cosmogonic one: If the earth and Venus formed close together, why did one receive so much water and the other so little?

Eventually it should be possible to choose between these alternatives. If Venus once had water but has lost most of it, the water that remains should be enriched in deuterium because that heavy isotope of hydrogen is less readily lost from the atmosphere. Hydrogen in the solar wind, on the other hand, is depleted in deuterium, since that isotope is consumed faster in the sun's nuclear reactions. So far it has not been possible to measure the ratio of deuterium to hydrogen on Venus. In this matter as in others, recent studies of Venus have raised more questions than they have answered. With its deep, murky atmosphere and smooth surface, Venus remains a clouded crystal ball.

6

THE EARTH

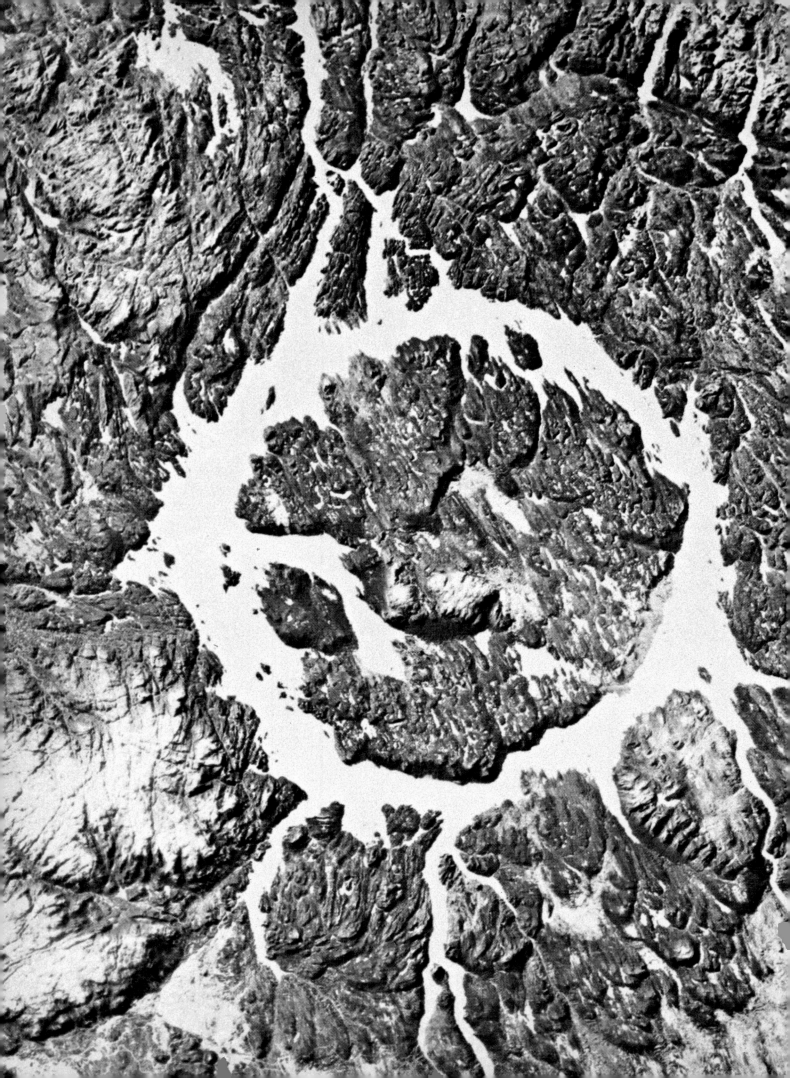

# The Earth

RAYMOND SIEVER

*The outstanding feature of our own planet is the dynamic activity of its atmosphere and its crust. Both have been substantially altered by the evolution of living organisms*

Apollo astronauts have said that the earth, with its blue water and white clouds, was by far the most inviting object they could see in the sky when they were on the moon. Their bias is understandable. They knew from intimate observation what this planet is like and could translate the sight of clouds, oceans and continents into everyday experience—of, say, a sea breeze blowing surf onto a sunny beach.

Probably the thing people like most about the earth, even if they have not put the thought into words, is its pattern of constant movement. On the earth stillness is remarkable for its rarity. Motion extends from the constant shifting of grains in a sand dune and the movement of bacteria and all other forms of life to the ponderous motions of the entire earth as it vibrates during and after an earthquake.

This planet is active. Indeed, it has been active for 4.6 billion years, and it shows no signs of calming down. The earth's atmosphere, oceans, thin crust and deep interior have been in motion since they were formed. Life has been an integral part of the surface for at least four-fifths of the planet's history.

As a consequence of its steady activity over this long span of time the earth has evolved through a series of quite different stages, maintaining during the entire time a state of dynamic equilibri-

um. The balance involves an exchange of matter and energy between the interior, the surface, the atmosphere and the oceans. It also involves sharing the radiation of the sun with the other members of the solar system. The study of geology, aided by work in geochemistry, geophysics and paleontology, has shown how the earth's surface skin has evolved. That knowledge, coupled with a firm theory about the constitution of the earth's interior and certain hypotheses on how the interior moves, provides a means for constructing a theory of how the planet evolved.

The article by A. G. W. Cameron [see "The Origin and Evolution of the Solar System," page 15] describes the origin of the earth and the other planets through the condensation of particular regions of the solar nebula. The original composition of the nebula and its later composition are deduced from the composition of the earth's rocks, of rocks brought back from the moon, of meteorites and of the atmospheres of the earth, Mars, Venus and Jupiter.

The theory of the earth's growth most favored by people who have studied the subject infers a gradual condensation and accretion of a solid planet as it swept up enormous quantities of small particles from the nebular disk that gave rise to the present solar system. As the

planet grew it began to heat up as a result of the combined effect of the gravitational infall of its mass, the impact of meteorites and the heat from the radioactive decay of uranium, thorium and potassium. (Although potassium is not normally regarded as being radioactive, .01 percent of the element on the earth is the radioactive isotope potassium 40.) Eventually the interior became molten. The consequence of the melting was what has been called the iron catastrophe, involving a vast reorganization of the entire body of the planet. Molten drops of iron and associated elements sank to the center of the earth and there formed a molten core that remains largely liquid today.

As the heavy metals sank to the core the lighter "slag" floated to the top—to the outer layers that are now termed the upper mantle and the crust. Accompanying the rise of the lighter elements, such as aluminum and silicon and two of the alkali metals, sodium and potassium, were the radioactive heavy elements uranium and thorium. The explanation for the rise of these heavy elements lies in the way atoms of uranium and thorium form crystalline compounds. The size and chemical affinities of the atoms prevent them from being accommodated in the dense, tight crystal structures that are stable at the high pressures of the deep interior of the earth. Therefore the uranium and thorium atoms were "squeezed out" and forced to migrate upward to the region of the upper mantle and the crust, where they could fit easily into the more open crystalline structures of the silicates and oxides found in crustal rocks.

As the earth was differentiating into a core, a mantle and a crust, the material at the top was also splitting into different fractions. The lower parts of the crust are composed of basalt and gabbro,

CRATER ON THE CANADIAN SHIELD, seen in the satellite photograph on the opposite page, was formed more than 200 million years ago by the impact of a large meteorite. The impact scar, a ring-shaped formation about 60 kilometers (37 miles) in diameter, is filled today by Lake Manicouagan, a major reservoir in northeastern Quebec. The accumulation of snow on the lake ice enhances the visibility of the formation. The bedrock, a Precambrian anorthosite, was melted and shattered by the impact; a peak of shocked rock remains in the center of the formation. Unlike the impact craters on Mercury and the moon, most of the craters on the earth were long ago erased or buried by erosional or tectonic processes. Many craters on the Canadian Shield, however, were preserved under layers of sediment and reexposed by glacial action. In this area glaciation removed 100 meters of sedimentary cover.

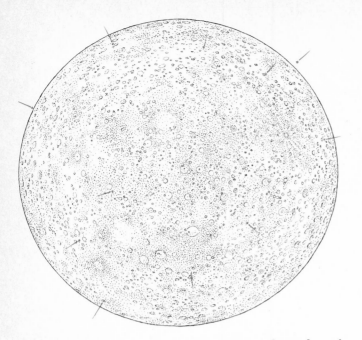

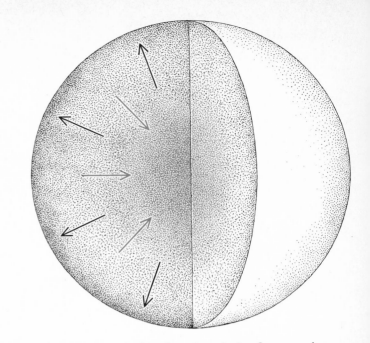

FOUR STAGES in the early evolution of the earth are shown in diagrams, beginning (*far left*) with the sphere that developed by accretion in the first few million years following the condensation of the solar nebula. The young planet is pockmarked by the infall of millions of planetesimals; its accretion atmosphere is rich in hydrogen and noble gases. Later (*left*), when some tens of millions of years have passed, a combination of gravitational compression, radioactive decay and impact heating produces melting and differentiation; heavy core and mantle materials sink inward and light crustal materials float outward. The solar wind sweeps away the accretion atmosphere; it is replaced by a primordial atmosphere rich in methane, ammonia and water. The Archaean era (*right*), be-

dark rocks containing calcium, magnesium and iron-rich compounds, mainly silicates. They were derived by partial melting and differentiation of the denser materials of the upper mantle. The basalt and the gabbro were themselves differentiated by fractional crystallization and partial melting and so, as lighter fluids, were pushed up through the crust. In the upper layers of the crust and at the surface they solidified to form the lighter igneous rocks such as granite, which are enriched in silicon, aluminum and potassium.

The question of how much of this "sweating out" process was completed at the early stage I am describing remains unresolved. Some geologists argue that a large amount, perhaps the bulk, of the granitic crust had already formed by this stage. Others cite the possibility that the process may hardly have begun even a billion years after the formation of the earth.

One result of the heating up of the interior was the inception of volcanic activity and mountain building. They contributed not only to the shaping of the surface but also to the immense change in the composition of the interior. During that time various gases, which had been locked in the materials of the planet when those materials originally accreted, began to find their way to the surface. They included carbon dioxide, methane, water and gases containing sulfur. The gases must have leaked to the surface in tremendous volume during the period of reorganization and differentiation. At the surface they stayed, since the earth's gravity was strong enough to prevent all but the lightest elements (hydrogen and helium) from escaping into space. The temperature at the time must have been low enough to allow the condensation of water. Dissolved in that water, the other gases combined chemically with elements such as calcium and magnesium, which were leached from surface rocks as rains began to weather them. If the temperature had been higher, the effect of the dense atmosphere with its large content of carbon dioxide would have been to institute the kind of "greenhouse effect" that seems to have arisen on Venus, producing that planet's hot, cloudy atmosphere [see "Venus," by Andrew and Louise Young, page 49].

As the surface of the earth cooled and the oceans formed from the condensation of water, the processes of erosion by wind and water began to operate in much the same way that they do today. Liquid water became the dominant mode for transporting and redistributing the debris of eroding mountains. The river systems of the surface are the visible traces of the networks that carry eroded material to the oceans, where much of it accumulates as aprons of sediment along the continental shelves and continental rises. The rest of the sediment is spread as thin layers over the ocean deeps by slow settling and the motions of turbidity currents.

A number of thoughtful geochemists and geophysicists have speculated on a somewhat different chain of events leading to the early accretion of the earth from the condensing solar nebula. According to these views, the earth and the other planets are the products of a gradual condensation of the solar nebula during which certain of the heavy elements, mainly iron, crystallized first, while the lighter fractions of the nebula were still in gaseous form. In that process the core of the accreting planet would have been iron-rich in the first place, with successively lighter fractions, roughly corresponding to the order of their crystallization from a gas, being accreted on the outside as the planet grew.

Whatever the mechanism of accretion, the story of the earth's later evolution (after the first billion years) is largely told by the record contained in the rocks of the crust. What they reveal is best told in terms of a geologic "clock" that began to run in Precambrian times. The oldest rocks now known are a series of metamorphosed sedimentary and volcanic rocks, which from their content of radioactive elements can be given an age of about 3.7 billion years. They are

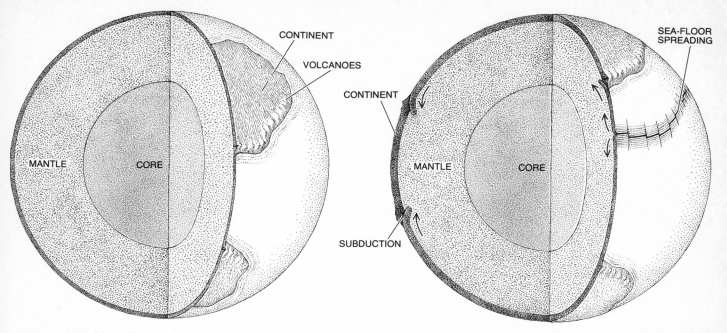

tween 3.7 and 2.2 billion years ago, is marked by cooling. As atmospheric water and gases condense, the oceans appear; the earliest continents arise and volcanic activity is intense. Chemical reactions between water, gases and crustal materials give rise to sediments and solutions. The Proterozoic era (*far right*) follows the Archaean era. It harbors a few scant traces of plant and animal life. It ends some 600 million years ago as the Paleozoic era, with its rich fossil record, begins. During that era's 1.5-billion-year span crustal cooling and thickening continue. Crustal plates begin migrating as such mechanisms as sea-floor spreading and subduction become established. Long before the end of the Proterozoic era both the interior and the surface of the planet have assumed their modern character.

much older than most of the very old rocks from the interval of time known to geologists as the Archaean. The rocks of that time are roughly defined as being older than 2.2 to 2.8 billion years. (The age of the boundary with younger eras varies in different parts of the earth's ancient rock terrains.) Much of the rock record is fragmentary, but it is tangible, and one no longer has to rely solely on plausible theory.

The Archaean rocks appear to be somewhat different from the rocks of all succeeding eras in the sense that certain rock types are abundant almost to the exclusion of many other types commonly found later. Archaean rock series tend to be dominated by basalts and andesites, which are volcanic rocks rich in iron and magnesium, deficient in sodium and potassium and relatively low in silica. The sandstones and shales of Archaean time were derived by the weathering and reworking of those volcanic rocks. Large bodies of granite—rocks richer in alkalis and silica—are absent. Such a skewed composition with respect to later rocks suggests that the sweating out of granitic rocks by fractional crystallization and partial melting of less silicic rocks was not as advanced as it became later.

The Archaean rocks also suggest that the tectonic style of the time, that is, the mountain-making activity by which the surface was shaped, differed from the pattern of today. The present theory of plate tectonics visualizes large plates of the lithosphere (which includes the crust and part of the upper mantle) moving laterally over the asthenosphere (a hot, plastic and perhaps partly molten layer of the mantle). The driving force is movement in the mantle, although the precise nature of that movement is uncertain. The geologic activity of earthquakes, volcanoes and mountain building is concentrated along the plate boundaries.

Granted that the Archaean rocks are widely dispersed and offer only a few bits of information, the study of the oldest terrains of Archaean areas in Canada and areas of similar age in Africa and Scandinavia does not suggest mountain building along the boundaries of large plates. It does suggest patterns of intense deformation along the boundaries of irregular areas of far smaller extent than plates. Many geologists suspect that the Archaean was a time of very thin lithospheric crust, extensive volcanism and some jostling movement of many small, thin "platelets," with "sutures," or crumpled deformational belts, welding them together.

Although the Archaean era differed markedly from the present in tectonic style and in the average composition of its volcanic rocks, it was the same as the present in all essential processes of erosion and sedimentation on the surface.

All the earmarks of weathering, mechanical breakup of rocks, transportation by rivers and sedimentation in regions where the crust gradually subsided and allowed great thicknesses of sediment to accumulate are found in Archaean sediments, as was pointed out more than 30 years ago by Francis J. Pettijohn of Johns Hopkins University, who was studying early Precambrian sedimentary rocks in the region of Lake Superior. Looking at those sandstones, shales and conglomerates, it is difficult to see any significant difference between them and more recent ones, all being the hardened equivalents of the gravels, sands and muds of today.

The erosion and chemical decay of rocks today are profoundly affected by the presence of land plants. It is known, however, that the higher (vascular) land plants did not evolve until two billion years after Archaean time, that is, in the middle of the Paleozoic era. Perhaps before the plants evolved, lower forms of life existed on the land, as they surely did in the sea.

Evidence of algal life late in the Precambrian era was obtained some years ago when the paleobotanist Elso S. Barghoorn of Harvard University, working with the late Stanley A. Tyler, a sedimentologist at the University of Wisconsin, discovered microscopic remains of algal organisms in the Gunflint

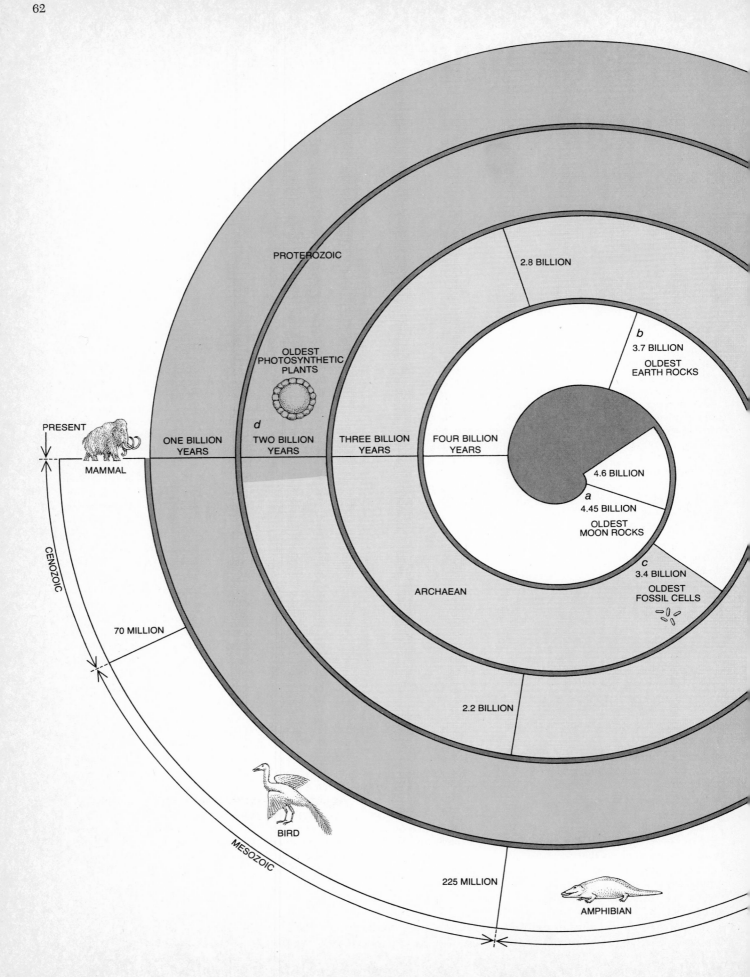

PROTEROZOIC

2.8 BILLION

*b*
3.7 BILLION
OLDEST
EARTH ROCKS

OLDEST
PHOTOSYNTHETIC
PLANTS

PRESENT

*d*

ONE BILLION
YEARS

TWO BILLION
YEARS

THREE BILLION
YEARS

FOUR BILLION
YEARS

4.6 BILLION

MAMMAL

*a*
4.45 BILLION
OLDEST
MOON ROCKS

CENOZOIC

*c*
3.4 BILLION
OLDEST
FOSSIL CELLS

ARCHAEAN

70 MILLION

2.2 BILLION

BIRD

MESOZOIC

225 MILLION

AMPHIBIAN

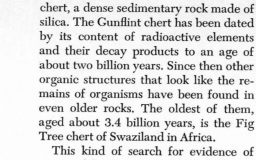

PRECAMBRIAN
INVERTEBRATES

600 MILLION

INVERTEBRATE

PALEOZOIC

BONY FISH

chert, a dense sedimentary rock made of silica. The Gunflint chert has been dated by its content of radioactive elements and their decay products to an age of about two billion years. Since then other organic structures that look like the remains of organisms have been found in even older rocks. The oldest of them, aged about 3.4 billion years, is the Fig Tree chert of Swaziland in Africa.

This kind of search for evidence of ancient life is a painstaking, laborious process. Thousands of rock specimens have to be sawed into ultrathin slices and then polished so that they can be studied under the light microscope and the electron microscope. Although organic carbon had been found in old rocks long before the discovery of the Gunflint and Fig Tree cherts, one could always hypothesize a variety of ingenious chemical mechanisms to account for it. The more recent evidence of distinctive forms of cellular life in ancient times is difficult to refute.

How life began on the earth is another story. It is the story of plausible chemical mechanisms that can be deduced from certain assumptions about the early chemical environment of the surface. One begins by inferring an early Archaean atmosphere (which had been built up by the escape of gas from the interior) that was dominated by water, methane and ammonia. Free oxygen was absent, since free oxygen is a product of life and not an antecedent to it. The atmosphere may also have contained appreciable quantities of carbon dioxide.

The existence and character of this atmosphere are related to the fact that the earth is smaller than Jupiter and larger than the moon. Jupiter was able to hold its hydrogen, which was by far the most abundant element in the solar nebula. The moon could not hold any of its gas.

In the earth's envelope of air and below it, in the surface waters of the sea and in large lakes, ultraviolet radiation from the sun was intense. The surface was not screened from the ultraviolet by a layer of ozone, as it is now, for want of the oxygen ($O_2$) from which ozone ($O_3$) is derived. The high energy of the ultraviolet radiation promoted the synthesis of a variety of organic compounds, for example amino acids. Perhaps many of these compounds were already here, since it is now known that a number of simple organic compounds are present in interstellar space.

The synthesis of transitory organic compounds, however, is not the same as making life. The next steps had to be the growth of large molecules and, before long, the growth of the nucleic acids that would eventually provide the genetic mechanism of reproduction so that cells could divide and give rise to new cells like themselves.

One cannot be sure of the range of chemical environments that will support life. (The uncertainty may be diminished by the U.S. spacecraft scheduled to land on Mars next year.) All that is known now is that the earth supports life and that its life depends on the continuous existence of liquid water. At present the earth is the only planet known to satisfy that condition. The earth's continuous record of life for at least the past 3.5 billion years shows that liquid water has been available during all that time.

Once life evolved, it began to exert an important effect on the surface of the earth and the gaseous envelope surrounding it. In the Bitter Springs formation of central Australia, which is a little less than a billion years old, paleobotanists have found cellular algae showing many of the geometric characteristics of the blue-green algae that today, like all other photosynthetic plants, evolve oxygen as a waste product. By the end of the Proterozoic era, which lies between Archaean time and the beginning of the Paleozoic era, there must have been enough oxygen in the atmosphere to support the evolution of higher organisms. They were the metazoans—animal organisms having many cells with differentiated characteristics. All these organisms appear to need at least small

**SPIRAL "CLOCK"** shows the passage of 4.6 billion years of earth history; each revolution of the hand takes a billion years. Moving clockwise toward zero (the present), the hand passes the first significant datum at 4.5 billion years before the present (*a*); this is the age of the oldest moon rocks known. The first complete revolution of the hand brings it to the oldest sedimentary rocks known on the earth, Archaean-era strata in Greenland (*b*). Some 350 million years later (*c*) the hand passes the earliest of certain fossil-like microstructures found in rocks from Swaziland in Africa; these may represent the planet's first flora. Almost a revolution and a half more are required to carry the hand past the algalike plants, roughly two billion years old (*d*), that are found in Canadian chert. Only two more revolutions remain to go; the Archaean era lies behind and most of the Proterozoic lies ahead. Half a revolution more will pass by three more familiar divisions of geologic time: Paleozoic, Mesozoic and Cenozoic. Human history occupies a hair's breadth to the left of zero.

quantities of free oxygen for their biochemical processes.

Oxygen is not the only atmospheric gas that comes from life. Methane, for example, is present in minute quantities. Its source seems to be primarily the methane-producing bacteria that yield the abundant "marsh gas" over swamps. The atmosphere also contains other gases that are the distinctive product of life rather than of simpler nonbiological chemical reactions.

The Proterozoic era was a time when the world was populated by bacteria, algae and other primitive single-cell organisms, probably on land as well as in the sea. Their influence on surface processes is seen in the Proterozoic rocks. It is most distinctive in stromatolites: rock formations consisting of the limy secretions of mats of filamentous algae and the sediment trapped by them. Stromatolites are known today in places such as the Bahamas and Bermuda, where limestone is being laid down in tidal flats. Other evidence of Proterozoic life is found in the existence of a few coal beds formed from masses of carbonized algal remains.

If an observer had looked down on the earth from an artificial satellite in Proterozoic time, he would have described the surface in much the same way that an observer similarly situated would describe it today. Only a sensor that could determine the chemical composition of the atmosphere would reveal any differences. The evidence for the similarities is in the Proterozoic rocks, which are of the same types and abundances as the rocks from all later ages.

By late Proterozoic time the earth-moon system, after early instabilities, had settled down into much the same system we see today. Tides would have been somewhat higher than they are now, but they would not have been grossly different. At about the time the moon became a cold planet the long heating and differentiation of the earth's upper mantle and crust resulted in the extensive intrusion of great bodies of granitic rock and in patterns of mountain belts that suggest a plate-tectonic origin.

Another kind of evidence from both Proterozoic rocks and more recent ones reveals periodic reversals of the earth's magnetic poles during much of the earth's history. As a heated rock cools it is magnetized in the direction of the earth's magnetic field, and the pattern is frozen in when the rock solidifies. Similarly, certain sediments that contain magnetic particles will record the direction of the field at the time they were deposited. The causes of reversals lie in instabilities in the fluid motion of the core, which is the driving force that creates the earth's magnetic field.

The same kind of paleomagnetic evidence reveals what has been called polar wandering, although it is not that the North and South poles have moved but rather that the surface features of the earth have shifted in relation to the poles. The conclusion is reinforced by paleoclimatic evidence, that is, the geologic record of ancient climates, such as the occurrence of coal beds in polar regions and glacial deposits near the Equator. From this kind of information it appears that a major continent was at the South Pole in Proterozoic time and that continental drift was already established as a major process affecting paleogeography.

The rocks also record from this time a major glacial epoch, the first for which there is firm evidence. The evidence is

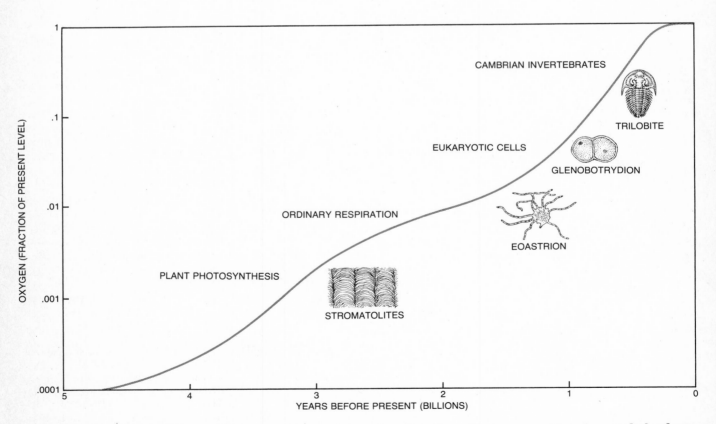

APPEARANCE OF OXYGEN in significant quantities in the earth's atmosphere, a late development, is an event that remains a subject of controversy. One hypothesis is shown in this semilogarithmic graph. The abscissa intervals are billions of years before present; increases in the oxygen supply, from trace amounts to the present quantity (about 20 percent of the atmosphere), are indicated on the ordinate. The process must have been gradual and must also have been related to an increase in the numbers of photosynthetic plants. The oxygen level may have risen to 10 percent of its present value a billion years ago but no evidence of oxygen-dependent animal life in any abundance appears until the time of a steep increase in the oxygen supply at the end of the Proterozoic era.

insufficient to reveal the details of that ice age—whether it was of the same extent as the recent (Pleistocene) ice ages and whether, like them, it had many episodes of glacial advance and retreat. One can only assume that the mechanisms postulated for the Pleistocene ice ages are general ones that are set in motion once a continental land mass lies at one of the poles and restricts the ability of the ocean and the atmosphere to distribute heat evenly around the globe. To an external observer at that time the earth would have looked a little like Mars, except that there were still oceans at the Equator. One of the interesting questions about the earth's glacial epochs is why the earth remained poised at a temperature distribution sufficiently low to give rise to large polar ice caps but not to a complete freeze-over.

Just as human history merges with prehistory, so the most recent 570 million years of the earth's history (starting with the Paleozoic era) connect with the nine-tenths of earlier evolution that were long thought to be a mystery. For more than a century the past 570 million years have been regarded as the geologically "known" period; it is therefore often called the Phanerozoic from the Greek *phaneros,* "to reveal." Although early geologists recognized that some Precambrian terrains were mappable by ordinary geologic methods, it was the absence of fossils having recognizable affinities with forms of the present that made it unknowable. The stratigraphic time scale, a marvelously detailed and precise clock, depends on the rapid evolutionary changes in higher forms of life that are recorded in the fossilized remains of corals, mollusks and thousands of other kinds of metazoan life.

Students of the earth's history never tire of marveling at the extraordinary speed of the coming of the metazoans. For between three and four billion years, almost its entire history, the earth was populated by single-cell life. Within at most a few hundred million years thereafter a fantastic diversity of invertebrate organisms appeared. All the major phyla of the animal kingdom became established quickly, and the vascular plants and the vertebrates soon followed.

Was all of this an accident, a favorable conjunction of continents, seas and environmental niches? Or was it the inevitable consequence of the buildup of oxygen in the earth's atmosphere by photosynthesizing algae? The best guess now is that it was the evolution of the atmosphere to near its present level of oxygen that stimulated biological inven-

tion. One such invention was animal shell, which served as armor to protect soft bodies from predators and as a base of attachment for muscles. Shells provide the basis for our understanding of the subsequent course of evolution of both the planet and its inhabitants. A paleobiological record based solely on the soft parts of organisms would provide only the dimmest outlines of the past.

The shells are more than markers of history: they influenced important changes in the dynamics of the earth's exterior. The oceans became populated with organisms that secreted calcium carbonate, calcium phosphate and silica in enormous quantities. Their remains were deposited as sediment, ultimately to become limestone, chert and phosphatic limestone or phosphate rock (a major source of agricultural fertilizer).

The more precise knowledge afforded by Paleozoic records enables geologists to trace the effects of continental drift. In particular it is possible to map more confidently the shape of the early Atlantic Ocean that lay between the European-African land mass and the Americas before the supercontinent of Pangaea was assembled at the close of the Paleozoic era. The assembly of Pangaea was one of the rare, special events of the later history of the earth, one of the important perturbations of the otherwise more or less evenly ordered evolution of the planet.

One of the major consequences of the assembly of Pangaea was the extinction of hundreds of species of invertebrates and the beginning of a wholesale change in the kinds and relative populations of the different animal and plant species. Most of the expanse of shallow shelf surrounding each continent disappeared as the continents collided, leaving only one narrow perimeter around the supercontinent. The shelves had harbored the most productive biological populations of the Paleozoic world. The geographic constriction and the concurrent climatic extremes, including the glaciation of parts of what are now Africa, Australia and South America, were enough to decimate many species. The survivors went on to found the new stocks of the post-Paleozoic world.

Pangaea rifted apart in the Triassic period (the earliest part of the Mesozoic era), and with that event and the following opening of the Atlantic Ocean and the drifting of the continents to their present position the story of the earth's physical evolution is largely told. The oldest parts of the ocean floor that are now preserved came into being at this

time, and so began a decipherable history of the world's oceans. It is read from the magnetic "stripes" and the fracture zones of the sea floor formed at mid-ocean ridges and rifting zones.

The new forms of life that evolved early in the Mesozoic era give the appearance of the modern world. Flowering plants appeared, and the lands became covered with the colors of the flowers and foliage of deciduous trees, the grasses and a great number of shrubs and flowers. In the sea new photosynthesizing algae, the diatoms, evolved; they are single-cell creatures secreting thin shells of silica. The diatoms became responsible for much of the primary photosynthetic production of organic matter in the sea.

At about the same time the calcareous foraminifera appeared. They are single-cell animals that live off the plants at the surface of the sea. Their shells of calcium carbonate, raining steadily to the bottom of the oceans, became the source of a new kind of deep-sea sediment, the foraminiferal oozes. The remains of these foraminifera became part of another detective story: the deduction of ancient sea temperatures, and thus of world climates, from the isotopic composition and external form of the shells. Both the shape of a shell and the relative proportions in it of the normal oxygen atom (oxygen 16) and the rare heavy isotope (oxygen 18) reflect the temperature of the water in which the animal lived. The temperature of the oceans as measured in this way has revealed an important climatic change.

Over most of the past 50 million years (during much of the Cenozoic era) the earth was cooling. This culminated in the past few million years in repeated glaciations. The more recent ones have been witnessed by and have affected the evolution of a new species: man. Already advanced on his course of evolution, man in his primitive cultures was displaced as the glaciers covered much of northern Europe, Asia and North America. In the short 10,000 years since the glaciers retreated to their present ice-cap size (probably a temporary retreat) man became the species that spread and occupied almost every environment of the surface of the planet. As he did so he became the latest of the biological populations to profoundly affect the course of the earth's history. He is only now becoming aware that some of his activities may alter the thin envelope of the atmosphere and the oceans and the fresh waters that make his existence possible.

7

THE MOON

# The Moon

JOHN A. WOOD

*Lacking an erosive atmosphere and geologically active outer layers, the earth's lifeless satellite has preserved a record of early events (but not the primordial events) in the history of the solar system*

No prospect could be more exciting to an earth scientist than to see an entire new planet suddenly opened up for study. The Apollo missions to the moon accomplished just that, although considerations of national prestige, not of science, had been the principal reason for going there. At a cost amounting to less than 5 percent of the total for the Apollo program, the scientific work associated with the program succeeded in transforming the moon from a cold distant circle of pallid whitish material into a real, if small, planet made of more or less familiar rocky substances in which a record of past epochs of geologic activity is preserved.

The scientific work involved many people and approaches. It is symbolized by the tools that were employed: microscopes, mass spectrometers, nuclear reactors, magnetometers and seismometers (the last left forever with their ear pressed against the surface of the moon, listening to its internal rumblings). The results of these studies can now be pieced together into a reasonably coherent picture of the composition and evolution of the moon. Much remains to be learned, but in broad outline the moon's life history has come to be understood almost as well as the earth's.

The present conception of the moon can perhaps be best summarized chronologically, by discussing the principal epochs or stages the moon appears to have passed through. Six stages are rec-

ognized at present: the origin of the moon, the separation of a crust, an early epoch of volcanism, a period of bombardment by massive planetesimals, a later epoch of volcanism and finally a decline of activity to the apparently quiescent stage of the present.

The origin of the moon is the stage about which the least has been learned. Few clues are offered by the lunar samples, because they have all turned out to be geologically processed materials—rocks whose compositions were established by igneous processes inside the moon. (The point is made by analogy with the properties of igneous rocks on the earth.) If the astronauts had been able to collect samples of the primitive substance of the moon that had been spared later transformation, much might have been learned from them about the formation of the moon. No such samples came to hand, however, probably because it is unlikely that any primitive material has survived the turbulent early history of the moon.

Certain things have been learned about the conditions of the moon's origin. First, the moon and the earth were formed in the same general region of the solar system. This conclusion is based on the isotopic composition of oxygen in the lunar samples, which is indistinguishable from the composition of terrestrial oxygen. The study of meteorites shows that the proportions of the iso-

topes oxygen 16, oxygen 17 and oxygen 18 vary measurably among rock samples derived from different parts of the solar system.

That observation does not much constrain hypotheses on the origin of the moon. It can still have been formed by fission from the earth, by capture intact from a nearby independent orbit or by accretion from small objects that once traveled in orbit around the earth. The oxygen data, however, do rule out the possibility that the moon was captured by the earth after having been formed far away, near Mercury or among the Jovian planets, for example, or outside the solar system altogether.

Second, when the lunar rocks are compared with terrestrial rocks or with meteorites, they are found to be systematically depleted in the more volatile chemical elements. The depletion can be seen in a comparison of elemental abundances in lunar and terrestrial basalts, to take one example [see illustration on page 71]. To be sure, basalts are secondary igneous rocks, not samples of planetary materials as they first accreted, but compositional differences between basalts from two planets should reflect differences in the composition of the more primitive planetary materials from which the basalts were derived.

The terrestrial planets (Mercury, Venus, the earth and Mars) are currently understood to have been formed by a process involving, first, the condensation of small mineral grains in the gaseous nebula that is thought to have surrounded the infant sun and, second, the mechanical accretion of the mineral grains into planets. As the initially hot nebula cooled, the most refractory minerals would have condensed first. Increasingly volatile compounds would have precipitated subsequently.

If condensation and accretion pro-

SURFACE OF THE MOON was photographed from an altitude of 118 kilometers by Alfred M. Worden, pilot of the command and service module of *Apollo 15*. His vehicle stayed in orbit around the moon while the other two astronauts in the crew were on the lunar surface. The cratering in this area, which is on the far side of the moon east of the crater Tsiolkovsky, is typical of the entire lunar surface and reflects a period of bombardment by planetesimals and meteoroids early in the evolution of the solar system. The crater in the center of the photograph is about 19 kilometers in diameter. It is younger than its neighbors, as is indicated by the bright "rays" of ejected material that overlie the nearby features.

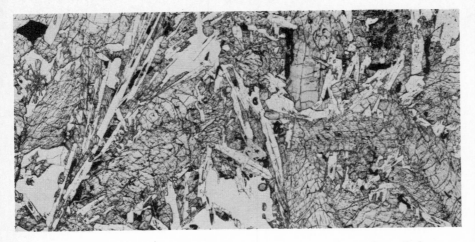

THREE TYPES OF LUNAR ROCK appear on this page in micrographs made by transmitted light at an enlargement of 12 diameters. This micrograph shows a thin section of basalt, which is the rock type found on the maria, or dark "seas" of the moon. The basalt displays a pattern of interlocking silicate and oxide minerals that crystallized when a flow of basalt cooled in Mare Imbrium. The rock specimen from which the thin section was cut was brought back to the earth by the astronauts of the *Apollo 15* mission in 1971.

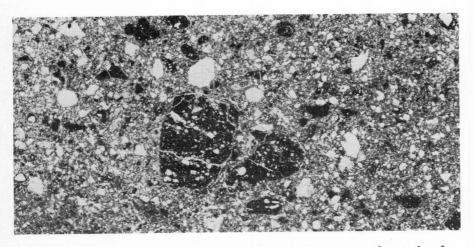

"KREEP" NORITE is a second type of lunar rock, one of the two types that are found on the light-colored highlands of the moon. The letters KREEP refer to the mineral's relatively high content of potassium (K), rare-earth elements (REE) and phosphorus (P). This specimen is a breccia, a conglomeration of mineral and rock fragments that resulted from the shattering of the precursor material in the bombardment by massive meteoroids early in moon's history. Rock from which section came was brought back by *Apollo 14* astronauts.

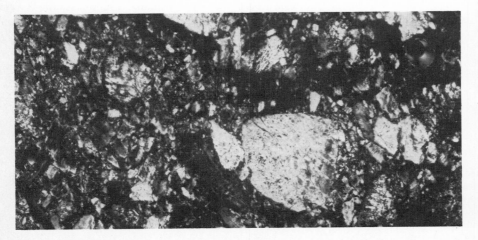

ANORTHOSITIC ROCK is by far the most abundant type of rock on the moon. This specimen, a breccia consisting largely of plagioclase fragments, was brought back by *Apollo 16* astronauts. For this micrograph polarizing filters were used to enhance the contrast.

ceeded simultaneously, circumstances could have conspired to deliver variable proportions of high-temperature and low-temperature condensates to the different planets. For example, the objects that began accreting first might be expected to capture the greater part of the early (high-temperature) condensates. The discovery that the moon and the earth contain quite different proportions of high-temperature and low-temperature elements makes it appear that such fractionation did occur when the planets were formed.

As for the actual mechanics of the formation of the moon as a satellite of the earth, the three possibilities I have mentioned, which were recognized long before the Apollo program, must still be considered. In the opinion of many lunar scientists, however, some variant of the model envisioning accretion in orbit around the earth is the most likely to be correct. The intact-capture model presents difficulties concerning dynamics. They can be summarized most simply by noting that the orbit of one object around another is symmetrical about the line between them at the time of their closest approach. If the moon approached the earth from some distant point in the solar system, it would, after its closest approach, recede to a similarly great distance along an orbital path nearly symmetrical with its approach trajectory. Unless special assumptions are made about a mechanism to slow the moon down while it was in the vicinity of the earth, it would not have been captured.

The idea that the moon fissioned from the early earth also presents difficulties. A very high rate of spin at the beginning must be assumed. After the rupture the earth-moon system would be left with a quantity of spin (angular momentum in the physicist's terms) twice as large as the quantity the system now has. Since angular momentum is ordinarily conserved in dynamic systems, special assumptions have to be made to provide a mechanism that would have de-spun the earth-moon system after fission.

If it seems preferable to adopt the hypothesis that the moon accreted as it was in orbit around the earth, where are the accreting particles to have come from? They probably condensed in the solar system at large and were subsequently captured in orbit around the earth. It is easier to capture a large number of small objects than it is to capture a large, preformed moon.

Several natural mechanisms would have acted to slow down small particles so that they could be captured by the earth. One is gas drag. The nebular gases would have resisted and slowed the motion of tiny particles but not of an object as massive as the moon. Another mechanism is collisions between particles. Approaching the earth from all directions, some particles would have passed around it in a clockwise direction and others would have gone counterclockwise. Collisions between members of these two populations would have slowed their forward velocities, guaranteeing that the particles could not recede from the earth. Collisions would have continued until all the particles either had lost enough velocity to fall onto the earth or had been bumped into orbits that were nearly circular, lay in a common plane and had the same sense of rotation.

It has been estimated that the moon could accrete from such a disk of particles in as little as 1,000 years. Whether or not it really formed as rapidly as that is unclear. If small particles continued to be captured into orbit around the earth from the solar system at large over a much longer period of time, the addition of particles to the moon would of course be correspondingly protracted. It is also unclear why the accreting moon would have captured more of the early (high-temperature) condensate particles than the nearby earth.

In order to discuss the internal evolution of the moon once it had formed, it is necessary to take stock of the types of rock found on the moon. Although a great variety of rock types were collected by the Apollo astronauts, nearly all the specimens can be put in one or another of three categories: mare basalt; "KREEP" norite, which is named for its unusually high content of potassium (K), rare-earth elements (REE) and phosphorus (P), and the anorthositic group. The mare basalt constitutes the substance of the dark and relatively smooth lunar maria, or "seas," and the other two rock types make up the rugged, light-colored highlands.

None of the three can be considered by any stretch of the imagination to be samples of primordial planetary material. Their elemental abundances bear little resemblance to the abundance pattern of metallic elements in the atmosphere of the sun or in chondritic meteorites, which are thought to be samples of primitive planetary material. On the other hand, all three do resemble classes of igneous rocks found on the earth. It is clear that igneous processes in the moon established the composition of the three categories of lunar rock.

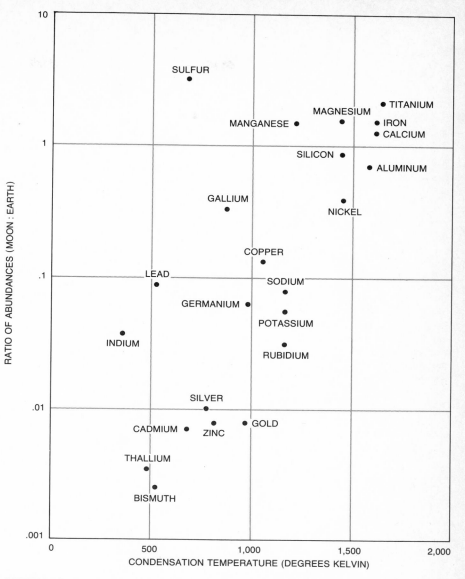

**COMPARISON OF BASALTS** from the moon and the earth shows that the lunar rocks are systematically depleted in the volatile elements that would have condensed at relatively low temperatures when the solar system formed from a gaseous nebula. The chart shows for each element the ratio of its abundance in the lunar basalts to its abundance in the terrestrial basalts. The contrast suggests a process of fractionation that delivered more of the early, high-temperature condensates to the objects in the solar system that accreted first.

Experience with igneous-rock systems on the earth makes it possible to recognize two fundamentally different igneous processes that fractionate rock chemically to produce new and different compositions. One is partial melting. To appreciate that concept the reader should refer to the phase diagram at the top of the next page, which illustrates the behavior with changing temperature of a material whose composition can be expressed in terms of the proportions of two end-member compounds. (It is therefore a binary system.) Phase diagrams of this kind are established by programs of careful experimentation, in which charges having various compositions simulating rocks are put in thick-walled vessels and taken to temperatures and pressures appropriate to the interior of a planet. The important point to be gained by the inspection of this particular diagram is that as the temperature is increased in a material whose composition can be represented in·terms of a mixture of diopside and anorthite, the composition of the first liquid to appear is the same no matter what the proportions of diopside and anorthite are in the starting material.

Actual rocky materials in the earth and the moon are too complex to be represented by a binary diagram; often their phase diagrams involve three or more end-member compounds, and sometimes they cannot be adequately

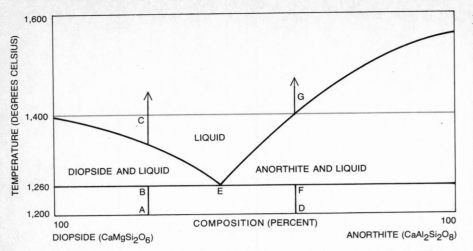

**PARTIAL MELTING** as a process that chemically fractionates rock to produce different compositions is described in this phase diagram. A material composed of the minerals diopside and anorthite in the proportions of 76 percent and 24 percent respectively remains solid as the temperature is raised (*AB*), until at 1,260 degrees Celsius (*B*) a melt of composition *E* appears. The melting is sufficient to eliminate solid anorthite. When the temperature rises further, the composition of the melt migrates (*EC*) as more and more residual diopside is melted. At *C* the last solid residue disappears. An equivalent thing happens if a solid containing different proportions of the minerals is heated, an example being 40 percent diopside and 60 percent anorthite (*DF, EG*). The key point is that for all starting compositions in this binary system the first liquid to appear has the same composition (*E*).

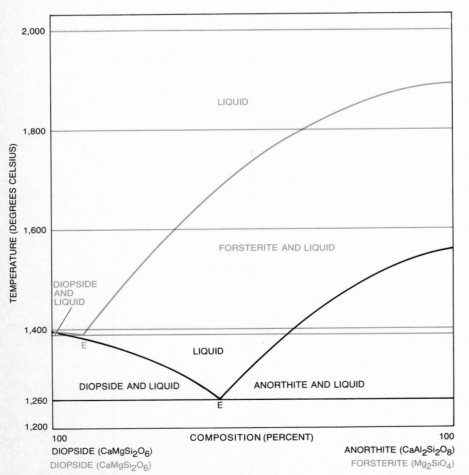

**PHASE DIAGRAMS** for two mineral compounds are compared. If a system is definable in terms of different end-member compounds, such as diopside and forsterite (*color*) instead of diopside and anorthite (*black*), a different low-melting liquid (*E*) is produced when the temperature is raised. Again, however, the composition of this low-melting liquid remains the same regardless of proportions of the end-member compounds in the starting material.

portrayed on two-dimensional paper. Nonetheless, the point made above remains valid: The first melt to appear when a mass of rocky material is heated has a characteristic composition.

When melting begins inside the earth, it appears that usually the early liquid does not remain long in the company of the residual solid material. Instead it tends to rise through the crust, frequently erupting onto the surface as lava. The composition of a terrestrial lava is often found to correspond to the low-melting composition for some particular phase diagram. The end-member compounds that define the diagram and the pressure regime for which it is valid can reveal the depth from which the lava was derived and what materials were present in the parent rock that gave rise to it. Thus the nature of planetary interiors can be probed by determining lava compositions and relating them to phase diagrams.

Returning to the trinity of lunar rock types, it turns out that mare basalt and KREEP norite have characteristic low-melting compositions and therefore were produced by partial melting in the interior of the moon. They represent later acts of lunar evolution and will be discussed below. The anorthositic rocks, however, have compositions far removed from low-melting liquids. These rocks must have been produced by the second type of igneous process, which is crystal fractionation.

When an igneous melt begins to crystallize, the crystals that form generally do not have the same specific gravity as the residual liquid does. If crystallization is not too rapid, dense crystals tend to sink to the floor of the cooling system and light crystals (if there are any) tend to float to the top. Layers can accumulate in which one mineral is greatly concentrated. The earth's crust contains layered igneous structures in which crystal fractionation has obviously occurred.

Anorthositic rocks are characterized by a superabundance of one mineral: plagioclase feldspar ($CaAl_2Si_2O_8$). It is easy to picture the formation of this type of rock by crystal fractionation in pools of lava on the moon. The difficulty is that the moon has vast amounts of anorthositic rock, which is overwhelmingly the most abundant of the three classes of lunar rocks. Apparently the entire crust of the moon, to a depth of from 50 to 100 kilometers and over the entire surface, is anorthositic. Local fractionation in lava pools would be totally inadequate to manufacture such an amount of anorthositic rock.

How can it be said that the moon has

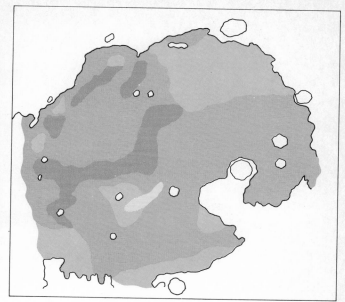

SUCCESSIVE LAVA FLOWS are evident in Mare Imbrium. They can be recognized by the different amounts of cratering they have undergone. At left Mare Imbrium appears in a Lick Observatory photograph that provides orientation for the map at right but is at too small a scale to show the density of cratering. Darkest color on map represents the youngest lava flow, which is the one least affected by cratering. Lightest color represents oldest flow. Second-youngest and most extensive flow erupted 3.3 billion years ago.

so much of this material? After all, the Apollo astronauts only scratched the surface at six points on the near side of the moon. (In addition the unmanned Russian spacecraft *Luna 16* and *Luna 20* sampled two points, and in 1968 the U.S. vehicle *Surveyor 7*, an unmanned lander, remotely analyzed the soil at a single point in the southern highlands.)

One can generalize about the composition of the crust to a considerable depth because, although the astronauts could not dig deep, earlier cratering impacts on the moon had done so. Huge basin-forming impacts delved tens of kilometers deep and scattered the excavated debris on the surface. Much of the material collected by the astronauts must have had this origin, and the material is dominantly anorthositic. The moon could not have kept another rock type hidden under a thin veneer of anorthositic material; cratering activity would have stripped the veneer away long ago.

Seismic studies have provided an actual measure of the thickness of the lunar crust on the near side. All the Apollo missions left functioning geophysical stations on the moon. Most of the stations include passive seismometers, which continuously transmit a record of seismic disturbances to the earth. Seismic signals were generated during the Apollo missions by deliberately crashing spent spacecraft onto the moon at preselected points. On the way to the various geo-physical stations the shock waves generated by those impacts passed through several materials having a range of seismic velocities.

When a model of the moon's crustal structure is devised that is consistent with the travel times of seismic waves from all the various impacts to the several seismometers, it is found to involve a discontinuity in the physical properties of the rock at a depth of about 60 kilometers. Above that depth the seismic velocities are consistent with the velocities in anorthositic rock. Below 60 kilometers the seismic velocities are higher (about eight kilometers per second compared with about 6.5 kilometers per second above that level). The velocity below 60 kilometers is appropriate to a rock type of higher density. Presumably that is the material of the lunar mantle.

The moonwide extent of anorthositic rock has been confirmed by remote geochemical analysis performed by instruments carried by the command and service modules of *Apollo 15* and *Apollo 16*. (The command and service module is the vehicle in which the third astronaut remained, traveling in orbit around the moon while the other two astronauts explored the lunar surface.) The vehicle kept a battery of sensing instruments trained on the lunar terrain. Among them was a detector sensing secondary X rays being emitted by elements in the lunar soil as a result of stimulation by primary X rays from the sun. The detector discriminated between X rays emitted by magnesium, aluminum and silicon in the lunar soil.

Anorthositic rock has a characteristically high ratio of aluminum to silicon. The X-ray experiment showed that material of this kind constitutes the substance of most of the lunar highlands the detectors passed over. The finding provides a basis for confidence that the highlands in general are anorthositic in character, although one could wish to confirm the fact by studying samples collected on the far side of the moon and in the polar regions.

If anorthositic rock can be formed only by crystal fractionation from a cooling melt, and if the layer of anorthositic rock produced by this process is approximately 50 kilometers thick over the entire surface of the moon, one is forced to postulate the existence early in lunar history of a layer of magma on a heroic scale—a veritable ocean of white-hot melted rock that once covered the surface of the moon. The present lunar crust must have separated from the hellish bath as it cooled. It is questionable whether crystallizing plagioclase would have actively floated up to the surface of this rocky ocean, because the specific gravity of plagioclase and of residual magma would have been nearly the same. If, however, the other crystallizing minerals sank (they would have been rich in iron and magnesium and certainly denser than the liquid), the net effect would have been to concentrate plagioclase near the top of the system, thus accounting for the plagioclase-

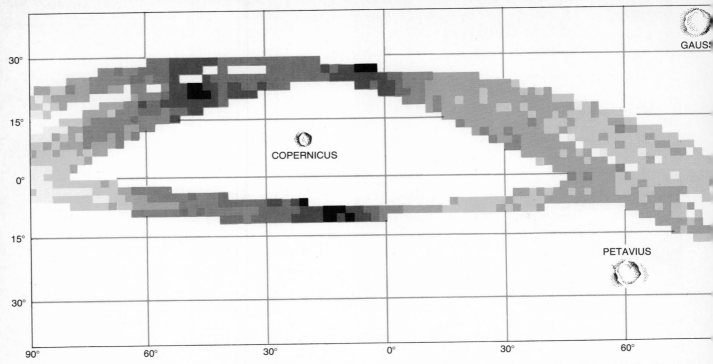

**NATURAL RADIOACTIVITY** in the lunar soil was detected by a gamma-ray spectrometer in the command and service module of *Apollo 15* and *Apollo 16*. The radioactivity is due to the decay of uranium, potassium 40 and thorium. The radioactivity in the most strongly colored regions of the map corresponds to the presence of about .3 percent potassium and seven parts per million of thorium.

rich (anorthositic) character of the present cool, solid crust of the moon.

The act of crust separation was the second stage of lunar history. All the terrestrial planets, including the earth, may have passed through a similar stage, although the record of the earth's primary crust has long since been obliter-

ated. The separation of the lunar crust probably occurred soon after the moon's formation.

That concept is not clearly supported by the ages of lunar anorthositic samples based on the decay of radioactive elements. Those ages are about four billion years, compared with an age of 4.6 bil-

lion years ascribed to the origin of the solar system and the planets in it. The radiometric ages probably do not, however, reflect the time when the highland rocks were formed. Later energetic events reset the radiometric clocks of the rocks.

The most compelling reason for plac-

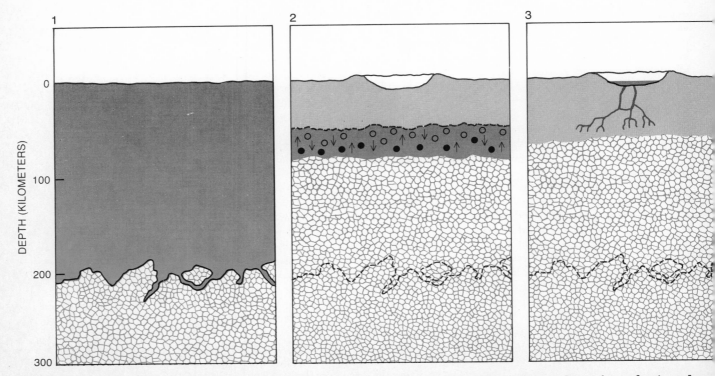

**INTERNAL EVOLUTION OF THE MOON** is summarized. The process involves melted rock (*dark color*), the lunar mantle (*light color*) and the crust (*gray*). The numbers refer to the six evolutionary stages discussed in the text. The stages are surface melting

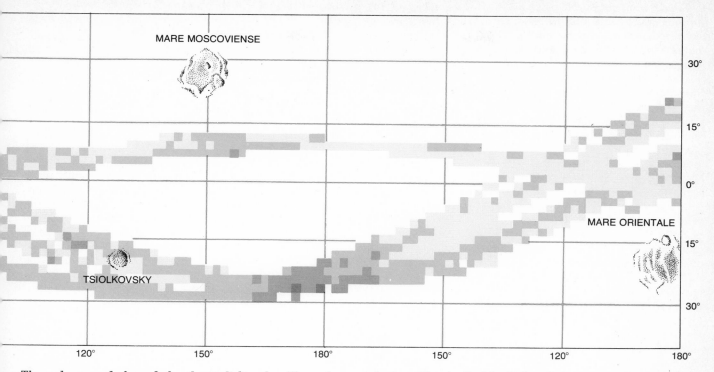

MARE MOSCOVIENSE

MARE ORIENTALE

TSIOLKOVSKY

The scale proceeds from dark color to light color. The regions that are portrayed in the lightest color represent concentrations of approximately .1 percent potassium and .3 part per million of thorium. The *Apollo 15* vehicle, which was in an inclined orbit, generated the sinusoidal pattern. The *Apollo 16* vehicle was in a nearly equatorial orbit and therefore generated a flatter pattern.

ing the separation of the lunar crust at the beginning is that it is easiest then to picture why the surface layers of the moon might have been molten. Planetary origin is likely to have been a violent process. Particles or subplanets accreting to an object having a substantial gravitational field would have arrived with considerable speed and kinetic energy. The kinetic energy would have been converted to heat on impact. If lunar accretion proceeded more rapidly than heat could be dissipated (the reader will recall that accretion could have been completed within 1,000 years), the heat resulting from the impacts would have been conserved and would have melted the outermost several hundred kilometers of the moon. It is hard to explain extensive surface melting at any later time; the normal state of affairs in a planet is for the surface, which is exposed to space, to be cool.

The third stage in the moon's history

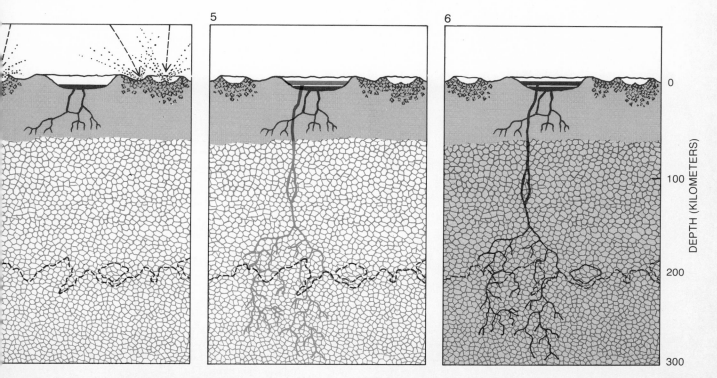

as moon formed (*1*), separation of a low-density crust by crystal fractionation (*2*), origin of KREEP norite by partial melting in the deep crust (*3*), intensive cratering (*4*), the origin of the mare basalts by partial melting in the upper mantle (*5*) and quiescence (*6*).

saw the appearance of the KREEP norite. This material's content of the major elements is not extraordinary, but its content of certain minor and trace elements such as potassium, phosphorus, barium, the rare earths, uranium and thorium is from 50 to 100 times higher than that of the lunar anorthositic rocks. As I have mentioned, KREEP norite has a low-melting composition and probably was produced by partial melting in the lunar interior. The appropriate phase diagram has as end-member components the same minerals that are found in lunar anorthositic rocks. The diagram is also valid at relatively low pressures. Apparently the most suitable site for the production of KREEP norite was in the anorthositic crust, after it had separated.

The minor and trace elements that are enriched in KREEP norite have in common rather large ions (atoms in the crystal structure that have gained or lost electrons). Such ions are not easily accommodated in the crystal structure of the major minerals in anorthositic rock. As a result these elements would have been among the last to solidify when the anorthositic system cooled, and they would be among the first to be remobilized if it were heated up again. The high concentration of large-ion elements in KREEP norite is therefore consistent with the hypothesis that the norite was produced by partial melting in anorthositic rocks. The concentration of large-ion elements also implies that only the first small percentage of liquid had been

sweated out of the parent rock before the liquid was separated and concentrated in bodies of KREEP-norite magma or lava.

Perhaps the most intriguing aspect of KREEP norite is its localized occurrence on the lunar surface. Gamma-ray detectors on the orbiting command and service modules sensed high concentrations of radioactive elements in the broad area of Mare Imbrium and Oceanus Procellarum, particularly at the points where light-colored highland terrain protrudes through the layer of mare basalt that generally covers the area. Evidently these highland regions consist of KREEP norite (which, as I have noted, is rich in uranium, thorium and potassium) rather than of anorthositic rock. It appears that at some time after the formation of the anorthositic crust of the moon, but before the eruption of the mare basalts that now blanket the area, a KREEP-norite lava flooded this one section of the moon.

The source of the heat needed to partly remelt anorthositic crustal material is not known, nor is the reason the eruption was concentrated in one area. It is tempting to postulate that one huge planetesimal impact provided the energy and the basin, but a crater as large as Oceanus Procellarum would have stripped the crust away entirely in some areas and promoted wholesale melting elsewhere. It would not have promoted the small degree of partial melting needed to generate KREEP norite. The origin of this

rock type remains one of the most important and puzzling questions of lunar science.

The fourth act of lunar history consisted of an epoch of impacts by major planetesimals on the surface of the moon. Our picture of the origin of the solar system involves the gradual assembly of small particles and accretions of particles into the present array of planets and satellites. The early solar system must have been an untidy place until the loose debris was swept up; meanwhile the debris bombarded the young planets ceaselessly. The crater-pocked surfaces of the lunar highlands bear witness to that early barrage of interplanetary projectiles. So does the character of the highland rocks, which are almost uniformly breccias: agglomerations of broken mineral and rock fragments. Pulverization from impacts has obliterated any textural evidence the old highland rocks once had of their origin in large-scale crystal fractionation or as lavas erupted onto the surface.

Major impacts not only fracture rock but also heat it. Severe heating of a rock has the effect of erasing any straight-forward isotopic record the rock has of the time when it was formed, that is, when it was endowed with its present chemical composition. The isotopic clock is reset, so to speak. That is the interpretation placed on the ages of lunar highland rocks. The age of four billion years at which they tend to cluster is not the

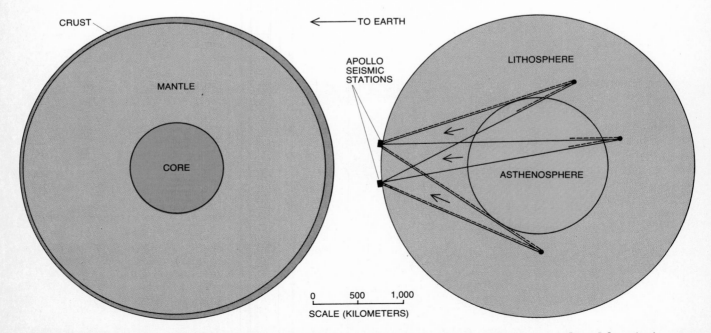

**PROPERTIES OF LUNAR INTERIOR** are portrayed. The distinction between crust, mantle and core (*left*) is based on differences in composition and specific gravity of rocks. The distinction between lithosphere and asthenosphere (*right*) is based on rigid-

ity (and therefore temperature) as indicated by seismic waves. Compressional waves (*solid lines*) from deep moonquakes (*dots*) pass through all material. Shear waves (*broken lines*) are attenuated by a nonrigid medium. They indicate a soft and warm zone.

time when they were formed; there is isotopic evidence (of a somewhat equivocal character) that both anorthositic and noritic lunar materials are older than four billion years. Some violent, high-temperature process reset their clocks four billion years ago.

The violent events in question were almost certainly the colossal impacts that excavated the huge circular mare basins in the lunar crust. Debris from the impacts blanketed much of the near side of the moon. Many of the highland samples collected must have been involved in these impacts and heated by them.

Apparently a new population of planetesimals was set loose in the solar system four billion years ago, resulting in a cataclysmic surge of cratering on the moon and other planets. It is also possible, however, that cratering activity was even more intense in the period prior to four billion years ago. Perhaps the early surface history of the moon was so violent that rock ages were constantly being erased, and not until four billion years ago did cratering activity decline to a point where rocks had a good chance of remaining undisturbed until the present day.

After the rain of planetesimals had abated, or perhaps while it was diminishing, the moon entered the fifth phase of its history. In this period vast floods of lava erupted on the moon's surface, flowing into the basins previously excavated by planetesimal impacts. There the lava solidified, forming the dark plains that appear to the unaided eye as smudges on the moon.

The mare lavas did not erupt in one pulse of volcanic activity. They continued to issue from the lunar interior for a period of almost a billion years. As a result the mare surfaces are a complicated patchwork of overlapping lava flows [see *illustration on page* 73].

Mare basalts are variable in composition as well as age. The oldest samples collected tend to have the highest content of titanium. It appears that they were produced by partial melting at a depth of 150 kilometers or more and that the less titaniferous basalts were generated later at greater depths (240 kilometers or more).

The source of basalt seems to have deepened with time. The phenomenon can be explained by the fact that the originally hot surface must have cooled rapidly by the radiation of heat to space, whereas the interior grew increasingly warmer through the decay of radioactive elements. The net effect must have been to cause the peak temperature to migrate inward with time.

The final stage of lunar history is quiescence. By the time the mare basalts erupted the density of major planetesimals in the solar system had fallen so low that large impacts ceased to be an important contributor to surface activity. That is why the maria have maintained such a smooth appearance to the present day. (They look smooth from a distance; close up they are seen to be pocked with small craters, the effect of continuing bombardment by the lesser meteoroids that have always been vastly more abundant than major planetesimals in interplanetary space.)

Internal disturbances as well as external ones have diminished during the last chapter of lunar evolution. The outermost shell of relatively cool and therefore rigid rock—the lithosphere—grew thicker as heat continued to be lost from the surface. By the time the youngest of the mare basalts were erupted the lithosphere had grown sufficiently thick and strong to resist sagging under the weight of the basalt flows.

More or less plastic masses of rock tend to sink or rise into positions of buoyant equilibrium. Where such movement is prevented by the brute strength of cold, rigid rock, and dense masses are held at unnaturally high elevations, the masses cause irregularities in the gravity field of a planetary surface. Such irregularities are termed positive gravity anomalies. It was already known before the Apollo program that positive gravity anomalies were associated with the circular maria; the mass concentrations ("mascons") in those maria had revealed themselves by deflecting the trajectories of orbiting spacecraft.

By now the lunar lithosphere has grown to a thickness of about 1,000 kilometers. The depth of the transition from the rigid lithosphere to the innermost, plastic asthenosphere can be read from the behavior of the seismic waves generated by moonquakes in the deep lunar interior [see *illustration on page 76*]. The presence of such a mighty armored layer on the moon absolutely prohibits fracturing of the lithosphere, jostling of fragmental plates and the transport of lava from the asthenosphere to the surface—processes that occur unceasingly on the earth, where the lithosphere is only from 70 to 150 kilometers thick. Vitality in the moon has retreated to a small central zone. To all intents and purposes that body—once the scene of thermal, chemical and mechanical activity on a gigantic scale—is dead.

8

MARS

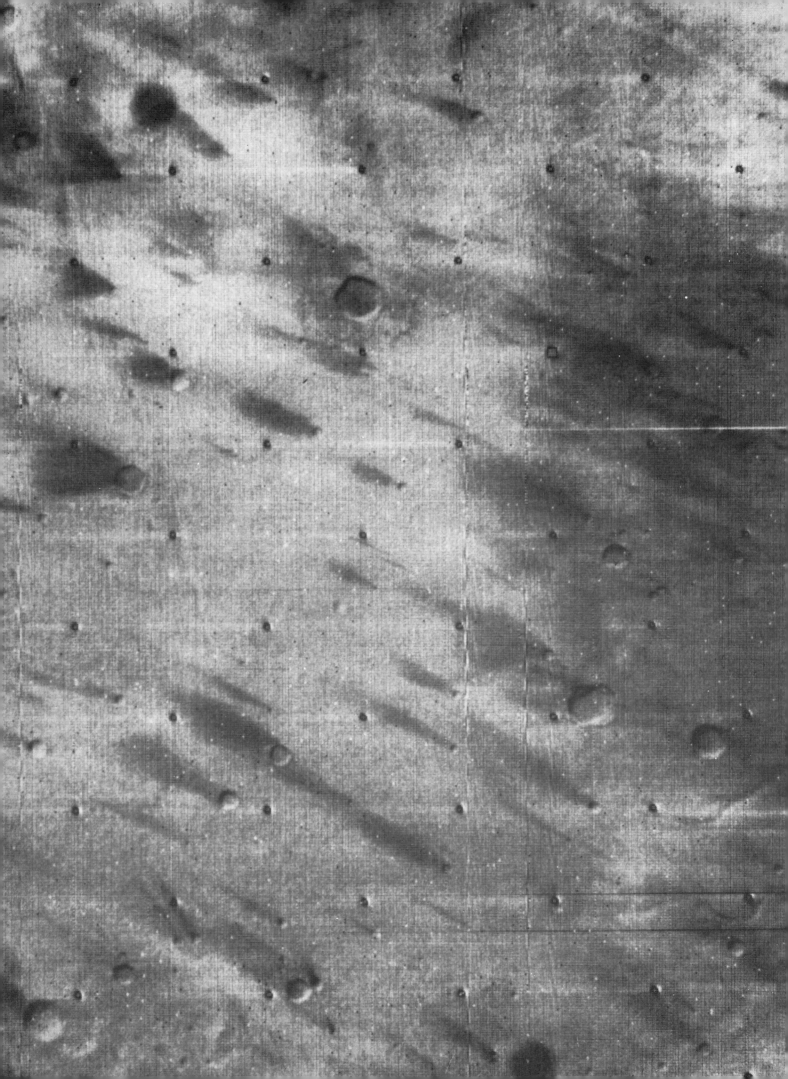

# Mars

JAMES B. POLLACK

*The first closeup photographs of it suggested that it was a cratered body as dead as the moon. Later pictures show a host of remarkable features indicative of a lively past*

No other series of spacecraft observations has resulted in such a profound revision of first impressions as the successive sets of pictures of Mars made by the Mariner probes that reconnoitered the planet in 1965, 1969 and 1971. The first set of 22 photographs, returned to the earth by *Mariner 4,* disappointed many because they revealed a drab and cratered planet rather like the moon. The 202 complete pictures returned four years later by *Mariner 6* and *Mariner 7* disclosed a few novel surface features but basically confirmed the impression that Mars was geologically dead. Although the cameras aboard the Mariners could not have provided evidence of Martian life, even if it exists, there was nothing in the pictures to raise the hopes of exobiologists. If Mars was as geologically dead as it appeared to be, it was probably biologically dead as well.

As it turned out, the first three Mariners sampled only a small and unrepresentative fraction of the Martian surface. When *Mariner 9* went into orbit around Mars in 1971, it revealed a planet with a spectacularly diverse topography, providing clear evidence of an active geological past. The pictures disclosed huge volcanic cones, one of them far larger than any on the earth, a system of gorges bigger than the Grand Canyon, vast sedimentary deposits in the polar regions and valleys that seem to have

been formed by running water. According to the new evidence, the climate of Mars may have been very different in the past from what it is today. It is now respectable to conjecture that if life got started under some earlier and more favorable climatic regime, it may have persisted under conditions that now seem hostile. The exobiologists may have their answer a little less than a year from now, when, if all goes well, the first of two unmanned Viking landers reaches the surface of Mars next July.

Many scholars of the 17th, 18th and 19th centuries believed that the climates of Mars and Venus were hospitable for life and even that intelligent beings populated these planets. It seemed that Venus might be a bit too warm and Mars a bit too cool, but temperature alone did not seem an insurmountable obstacle. It was known early in the 19th century that Mars rotates on its axis once in almost exactly 24 hours and that its axis is inclined to the plane of its orbit about 24 degrees (almost exactly the same as the earth's). Therefore it was reasoned that Mars should go through an earthlike sequence of seasons in its annual 687-day journey around the sun. Given that Mars has a diameter only about half as large as the earth's and a surface gravity only about four-tenths as strong, it seemed to early astronomers that Mars's atmosphere should be thin-

ner than the earth's, but estimates varied widely and the composition of the atmosphere was unknown.

Successive generations of observers made countless maps of Mars, plotting and replotting the locations of variable surface features to which they assigned such fanciful names as *mare* (sea), *sinus* (bay or gulf), *lacus* (lake), *lucus* (grove or wood), *fretum* (strait or channel) and *palus* (swamp). They also argued the reality of Martian canals well into this century.

By the time the first space vehicles were lofted toward Mars, however, astronomers generally agreed on a few specific observational features: that the Martian atmosphere contains clouds, that the white polar caps wax and wane with the seasons, that the planet's surface at various latitudes often shows seasonal changes in color and that the planet is periodically swept by huge dust storms. Spectrographic studies in the 1950's and 1960's showed that the Martian atmosphere is rich in carbon dioxide, very low in water vapor and apparently devoid of oxygen.

Even in the absence of samples from the surface of Mars a good deal can be inferred about its bulk composition. Its density, for example, is only 70 percent the density of the earth. Presumably Mars was formed in a region of the solar nebula that was cool enough to allow the condensation of compounds incorporating a wide variety of elements. The planet is probably well supplied with minerals containing combinations of magnesium, iron, silicon and oxygen (ferromagnesian silicates) and combinations of iron and sulfur (troilite). There may also be some free iron. Because Mars formed in a cooler region of the solar nebula than the earth did, however, more of its iron combined with other elements and less is in the form of

CRATERS ON MARS in this photograph made from the orbiting spacecraft *Mariner 9* have dark tails associated with them, all aligned in the same direction as the result of strong winds acting on the surface material of the planet. As the winds blow past the craters their speed increases on the craters' downwind side, removing bright dust particles from the ground and leaving the dark tails. The craters, like the craters on Mercury, Venus, the earth and the moon, are mostly due to heavy impacts early in the history of Mars, when the planet was bombarded by debris that had been left over from the formation of the solar system. The Martian craters, however, unlike the craters on Mercury and the moon, are modified and eroded by the scouring action of the high-speed winds in planet's atmosphere.

metallic iron. This difference may explain why Mars has a lower density than the earth. Temperature conditions also favored the formation of some water-bearing silicate materials. In addition to these more abundant materials Mars contains compounds incorporating small

amounts of the long-lived radioactive isotopes of uranium, thorium and potassium.

Only about a tenth as massive as the earth, Mars was evidently assembled rapidly, probably in less than 100,000 years. The more rapidly a planet is

formed, the larger is the fraction of the gravitational energy of formation it can store and the warmer it will be at the end of the accretion period. Moreover, the outer layers are heated to higher temperatures than the deeper layers. Current theoretical models of the earth

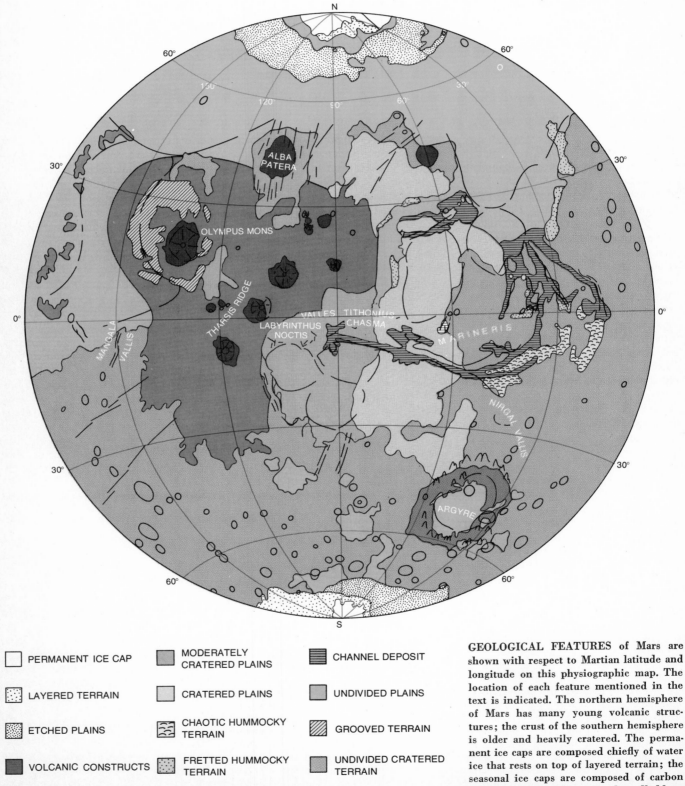

| | | |
|---|---|---|
| ▢ PERMANENT ICE CAP | ▨ MODERATELY CRATERED PLAINS | ▤ CHANNEL DEPOSIT |
| ▨ LAYERED TERRAIN | ▢ CRATERED PLAINS | ▢ UNDIVIDED PLAINS |
| ▨ ETCHED PLAINS | ▨ CHAOTIC HUMMOCKY TERRAIN | ▨ GROOVED TERRAIN |
| ▨ VOLCANIC CONSTRUCTS | ▨ FRETTED HUMMOCKY TERRAIN | ▨ UNDIVIDED CRATERED TERRAIN |
| ▨ VOLCANIC PLAINS | ▨ KNOBBY HUMMOCKY TERRAIN | ▨ MOUNTAINOUS TERRAIN |

**GEOLOGICAL FEATURES** of Mars are shown with respect to Martian latitude and longitude on this physiographic map. The location of each feature mentioned in the text is indicated. The northern hemisphere of Mars has many young volcanic structures; the crust of the southern hemisphere is older and heavily cratered. The permanent ice caps are composed chiefly of water ice that rests on top of layered terrain; the seasonal ice caps are composed of carbon dioxide ice. Layered terrain, also called laminated terrain, is a relatively young area

and the moon suggest that both had a high initial temperature and that both must have been formed quickly. The same seems to have been true of Mars. The two tiny satellites of Mars, Phobos and Deimos, seem to have been formed at the very outer edges of the ring of debris that gave rise to the main planet.

At the end of the accretion period Mars had a homogeneous composition and no atmosphere. The outer layers, which were molten right after formation, cooled and solidified to form a variety of rock types. The lighter rocks, which were enriched in silicon and aluminum and deficient in magnesium, floated to the top to form the crust of the planet, which is about 50 kilometers thick. Subsequent heating of the interior caused by the decay of the long-lived radioactive isotopes led to the planet's

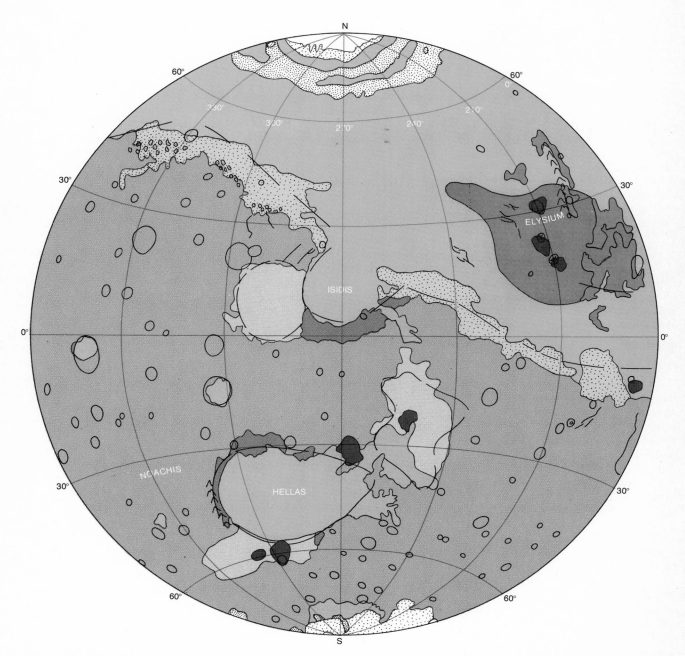

near the poles where successive blankets of dust and ice have been deposited. The etched plains are irregular, pitted, unlayered deposits that have been eroded by the wind. The volcanic structures are largely shields, domes or cones. The volcanic plains have few craters and many lobed scarps that seem to be the fronts of solidified lava flows. The moderately cratered plains lack volcanic structures. The cratered plains are the most densely cratered of the plains regions on Mars. They include ridges resembling the ones that are found on the maria ("seas") of the moon, and they also show some old, eroded volcanic features. The chaotic hummocky terrain is a depressed area consisting of disrupted and tilted blocks of the Martian crust. The fretted hummocky terrain is a lowland region with numerous isolated mesas bordered by cliffs. Knobby

hummocky terrain is an isolated region of knobs, each about 10 kilometers across. The channel deposits are the smooth floors of channels and canyons, probably composed of material deposited by either water, wind or landslide. The undivided plains have a generally scoured appearance, full of irregular ridges, scarps and channels, and are sparsely to moderately cratered. Grooved terrain is a line of mountains next to Olympus Mons that range between one kilometer and five kilometers wide and lie in an arc about 100 kilometers long. Undivided cratered terrain is a region of densely to moderately cratered uplands and is the most ancient part of the exposed surface of Mars. The mountainous terrain is the rugged area adjacent to a basin. It probably consists of material that was ejected by original impact that formed basin and was later eroded.

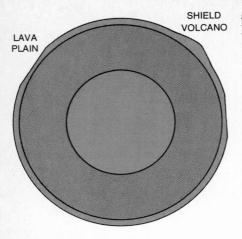

LAVA PLAIN

SHIELD VOLCANO

**CROSS SECTION OF MARS** shows how the planet may have differentiated after it accreted from the nebula that gave rise to the solar system. The crust (*gray*) is rich in aluminum and silicon and is deficient in magnesium. The decay of long-lived radioactive elements heated the interior of the planet until metallic iron and troilite (iron sulfide) melted and sank toward the center to become the core (*dark color*) and a mantle (*light color*) of ferromagnesian silicates was formed. Over the history of Mars some of the lighter molten rock has reached the surface, preferentially in the northern hemisphere, spilling over into huge lava plains and building additional volcanic structures.

further chemical differentiation. The interior became segregated according to the density of its constituent elements and minerals, and volatile gases were vented to the surface and formed an atmosphere. Two additional major stages of differentiation can be distinguished.

In the first stage the interior temperatures became high enough for mixtures of metallic iron and troilite to melt and sink to the center, forming a core of iron-rich material. Following this major rearrangement of material, Mars expanded slightly in size and its primordial surface was totally destroyed. Evidence that Mars does indeed have a dense core has been supplied by accurate measurements of the planet's gravitational field made by a succession of Mariner spacecraft.

The second stage of differentiation took place some time later than a billion years after the first, when temperatures became high enough in the outer layers of Mars to melt some of the rocky material in the mantle lying above the core. In the process some of the mantle was differentiated into a lighter rocky material enriched in silicon and aluminum and a denser material depleted in those two elements. The lighter material floated to the surface, creating plains

and volcanic structures. The volcanism has continued episodically up to the present, and the top of the zone of melting has shifted to progressively greater depths below the surface.

Interior melting also led to the generation of an atmosphere. When a rock is heated, it loses some of the elements that have been adsorbed on its surface, such as the noble gases neon and argon. Simultaneously some of the elements or molecules, such as water, that have been incorporated into the rock are released. Some of the gases set free in this way will recombine to form thermodynamically stable compounds. Carbon will combine with oxygen to form carbon dioxide and with hydrogen to form methane. The net effect of interior melting is the introduction of noble gases, water vapor, carbon dioxide, methane, nitrogen and hydrogen into the atmosphere. The ultimate composition of the atmosphere, however, is determined by the subsequent partial loss of some of these gases and by their chemical reactions with one another.

On Mars a sizable fraction of the out-gassed water vapor and significant amounts of the carbon dioxide either condensed as ices in the polar regions or were adsorbed on the surface of dust particles lying on the ground. Mars's gravitational field, like the earth's, is not strong enough to keep the lightest gas, hydrogen, from escaping from the top of the atmosphere.

The composition of the planet's early atmosphere depended sensitively on the concentration of hydrogen molecules in the lower atmosphere. If the outgassing rate was high and the escape rate of hydrogen was low, enough hydrogen may have resided in the atmosphere to reduce most of the carbon dioxide to methane and water vapor. At a later time in the planet's evolution the outgassing rate may have dropped sharply, with a concurrent drop in the hydrogen content of the atmosphere. In that event carbon dioxide would have emerged as the principal constituent of the atmosphere. Later I shall examine the consequences of an early reducing atmosphere for climatic change on Mars and for the possible evolution of life.

We now know from measurements made possible by the Mariners that the Martian atmosphere is very thin: its pressure at the surface is only about a two-hundredth of the pressure of the earth's atmosphere at the surface. Although the atmosphere is composed principally of carbon dioxide, it may also contain a surprisingly large amount of

argon, perhaps as much as 30 percent. Carbon monoxide and oxygen make up about .1 percent of the atmosphere; the amount of water vapor is variable but averages about .01 percent, and as much as one part per million of ozone is present, along with traces of atomic and molecular hydrogen.

A large fraction of the atmospheric carbon dioxide, possibly more than a sixth and as much as a fourth, condenses semiannually in the winter hemisphere, vaporizes during the spring and freezes again at the opposite pole. At their maximum size the seasonal polar caps reach down to a latitude of about 60 degrees. Large quantities of carbon dioxide may also be adsorbed on the surface of soil particles and possibly, although this now seems unlikely, be trapped in the permanent ice caps.

The amount of free water vapor in the Martian atmosphere is even more strictly limited by condensation and adsorption processes. Because water ice condenses out at a much higher temperature than carbon dioxide (190 degrees Kelvin compared with 150 degrees) it forms small but permanent polar caps in both hemispheres. In addition much water is probably trapped at all latitudes in the regolith, the granular material that lies above the bedrock. From infrared observations made from a high-altitude aircraft operated by the Ames Research Center of the National Aeronautics and Space Administration, James R. Houck of Cornell University and I have concluded that bound water is ubiquitous on Mars, constituting about 1 percent of the surface material. Such water may be present throughout the regolith, which is as much as a kilometer deep. Moreover, at latitudes higher than 45 degrees the ground below the surface is cold enough throughout the entire year for the atmospheric water vapor to condense and form subsurface ice layers, analogous to the permafrost layers found in the earth's polar regions.

Virtually all the water on Mars is tied up in the permanent polar caps, the permafrost and the regolith. If all the water remaining in the atmosphere were to condense, it would cover the surface of the planet with a liquid film only a hundredth of a millimeter deep. If the water in the polar caps were spread evenly over the surface, however, it would produce a layer 10 meters deep, and an equal volume of water may be tied up in the permafrost and the regolith.

If the amount of argon in the Martian atmosphere is as high as is suggested by recent Russian spacecraft observations, the amount of carbon dioxide and water vapor simultaneously outgassed from the planet's interior must have been very great. Argon constitutes about 1 percent of the earth's atmosphere, almost all of it manufactured by the decay of radioactive potassium in the earth's interior. If allowance is made for the amount of carbon dioxide now tied up in carbonate rocks and the amount of water in the oceans, one finds that about 2,000 times more carbon dioxide than argon and 40,000 times more water have been released from the earth's interior. Therefore if argon constitutes some tens of percent of the current Martian atmosphere, sizable quantities of carbon dioxide must be contained in the planet's regolith, and even larger amounts of water must be contained in the polar caps, the regolith and the subsurface regions than were given in the estimates above.

All the trace gases in the Martian atmosphere that have been detected so far—carbon monoxide, oxygen, ozone and hydrogen—are evidently formed by photochemical processes that begin with carbon dioxide and water vapor. These parent molecules are broken into molecular fragments by ultraviolet radiation from the sun. The fragments then proceed to react with one another and other parent molecules to give rise to the trace gases. The ultraviolet spectrometers aboard the Mariner spacecraft have directly observed the abundance of hydrogen atoms near the exosphere, or outer atmosphere, of Mars by means of the sunlight they scatter, making it possible to estimate the rate at which the hydrogen atoms are escaping. If the current rate of loss is typical of the rate throughout Mars's lifetime, it would imply the destruction of an amount of water sufficient to cover the planet to a depth of several meters.

It was originally thought that the oxygen left over from the breakup of water and the escape of hydrogen reacted with the crust of Mars to produce more highly oxidized minerals. Michael B. McElroy of Harvard University has proposed, however, that photochemical processes near the exosphere liberate oxygen atoms with velocities high enough for them to escape into space. This happens when some ionized molecular fragments that incorporate oxygen atoms combine with electrons to produce un-ionized molecular fragments. Some of the energy generated by the chemical transformation goes into energy of motion. Surprisingly, the amount of oxygen escaping as a result of this process appears to be comparable to the amount of hydrogen es-

caping. It seems likely that the amount of these two kinds of atoms that are being lost to space is in proportion to their relative abundance in water (two hydrogen atoms to one oxygen atom) and that there are no leftover products.

The photochemical mechanism for producing fast atoms that can escape may also be effective for atoms of nitrogen and carbon. This may explain the virtual absence of molecular nitrogen in the present Martian atmosphere, particularly if much of the outgassing of the atmospheric constituents came early in the planet's history. The carbon atoms lost to space by the same process over the lifetime of Mars would yield an amount of carbon dioxide equal to a large fraction of the current carbon dioxide content of the atmosphere.

The minute amounts of ozone found in the atmosphere of Mars are generated in the lower atmosphere by the chemical combination of the molecular and the atomic oxygen that are produced by photochemical reactions in the upper atmosphere. Ozone is found in sizable quantities only in the winter polar regions, where the air temperature is low enough to freeze out most of the water vapor. When water vapor is present, the products of its photodissociation react with ozone and lead to its rapid destruction. Studies of the Martian ozone cycle may yield a better understanding of the processes that control the amount of ozone in the earth's atmosphere, which acts as a shield in protecting life at the earth's surface from harmful solar ultraviolet radiation.

In spite of the tenuous nature of the Martian atmosphere, both condensation clouds and dust clouds are frequently observed in it. When the atmospheric carbon dioxide is condensing on the

GIANT SHIELD VOLCANO, Olympus Mons (formerly known as Nix Olympica), is some 600 kilometers in diameter. Its summit reaches a height of about 25 kilometers above the surrounding plain. The calderas, the smaller circular structures near the center, were at one time vents for the molten lava that helped to build up the shield and flowed down the flanks of the volcano, giving rise to the feathery texture radiating from the center. At the end of each period of the volcano's eruption, lava ebbed from under vents and the surface collapsed to form calderas. Cliff dropping to plain along perimeter of shield is two kilometers high.

ground in the cold polar regions of the winter hemisphere, clouds often develop. At middle latitudes in the same hemisphere the air temperature drops below 190 degrees K., and water vapor condenses into water-ice clouds. Closer to the pole the air temperatures approach the condensation point for carbon dioxide, 150 degrees K., and dry-ice clouds form. The cloud patterns often change from day to day.

The television cameras aboard *Mariner 9* photographed several interesting cloud patterns in the polar area. On a number of occasions clouds formed a remarkable wave pattern on the downwind side of a large crater. The flow of air over the elevated crater rim sets up a pattern of lee waves behind the crater. Such clouds are common in the lee of mountains on the earth. Where the air is rising it expands and cools, which leads to the generation of ice clouds at periodic intervals.

Localized condensation clouds are sometimes found on Mars at more equatorial latitudes. They are formed by the rising and cooling of air over highly elevated surface features, such as the large

volcanic mounds located in the Tharsis region or near it. The clouds usually become progressively brighter as the day advances from early morning to late afternoon. They are more frequent in summer, when the water content of the atmosphere is high. The infrared spectrograph aboard *Mariner 9* showed that the clouds absorb spectral wavelengths characteristic of water ice.

Dust storms confined to localized areas are frequent on Mars. In addition a great dust storm develops and covers a large portion of the planet every Martian year. The storm begins suddenly as a brilliant, elongated white feature several thousand kilometers in length. It generally originates at about 30 degrees south latitude and 320 degrees west longitude, in the northeastern Noachis region near its border with the great basin Hellas. The dust storm begins at the end of spring in the southern hemisphere, when Mars is near perihelion, its closest approach to the sun. For the first few days the central dust cloud expands very slowly. Thereafter it spreads rapidly, mostly in a westerly direction, and within a period of several weeks it complete-

ly girdles the planet. At the same time tongues of the storm develop and spread to higher and lower latitudes, local dust storms start at other places in the southern hemisphere and small amounts of dust are carried to other latitudes. Within a month after the start of the storm the dust cloud covers almost all the southern hemisphere. Sometimes, as in 1971, coincident with the arrival of *Mariner 9*, the dust storm spreads into the northern hemisphere and the entire planet becomes shrouded in a cover of dust. The dust storm subsequently decays as dust particles begin to settle out of the atmosphere, and within a month or several months the atmosphere again becomes transparent.

Wind velocities greater than 100 miles per hour are needed to set dust grains in motion at the surface and to lift tiny dust particles to high altitudes in the atmosphere. A great dust storm begins either because strong winds have developed on a global scale and become intensified in the Noachis region or because the large-scale winds are sufficiently calm to favor the development of intense local circular wind patterns, about

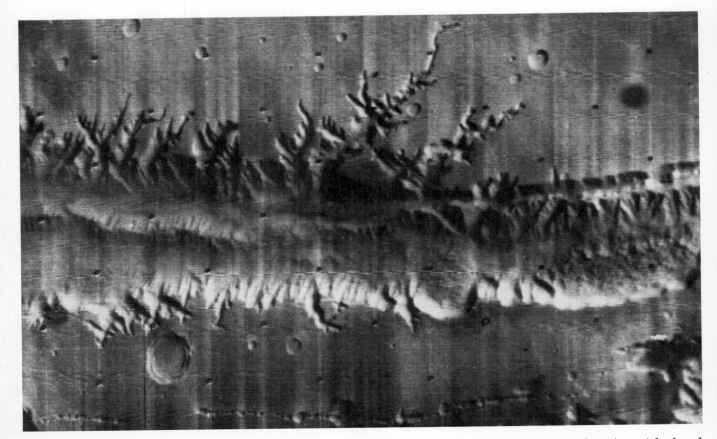

**ENORMOUS CANYON,** Tithonius Chasma, is about 75 kilometers wide and several kilometers deep. It is a part of the canyon complex Valles Marineris, which may have been created when the surface of Mars fractured, or when material below the surface was withdrawn, or when the subsurface ice melted (and in some cases gave rise to running water that carried away the material). The dendritic (branching) pattern on the southern (*upper*) bank and the intricate set of ridges on the plateau within the trough itself to the left may have been formed by running water. The line of pits that is visible along the northern highland border of the gorge at bottom of picture may be a canyon at an early stage of development. Entire Valles Marineris region is as long as the U.S. is wide.

a kilometer across, that are analogous to "dust devils" on the earth. Once dust is injected into the atmosphere a large organized circular wind system, about 1,000 kilometers in diameter, may develop that is analogous to a hurricane on the earth. In a terrestrial hurricane the heat needed to drive the strong winds is provided by the condensation of water vapor near the eye of the storm. In the hypothetical Martian hurricane the heat is provided by sunlight, which is absorbed by dust particles that then heat the surrounding gas. The buoyancy of the heated gas carries small dust particles to altitudes as high as 50 kilometers. As a result of the strong temperature contrast between the dust-heated air and the surrounding cooler air large, dust-raising winds are produced at the storm's periphery. In that way the storm grows. When dust covers the entire planet, the temperature differences are greatly diminished, the winds subside and the dust settles out of the atmosphere.

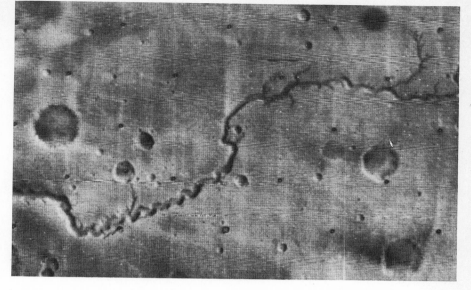

SINUOUS CHANNEL with an immature tributary system appears to have been created by running water or at least by some fluid less viscous than lava. The channel, Nirgal Vallis (located at 29 degrees south latitude, 40 degrees west longitude), is 1,000 kilometers long.

The surface of Mars is the product of a variety of processes that have created a wide diversity of geological landforms. These processes include bombardment of the surface by meteoroids, volcanic and tectonic activity, sapping (the removal of ground support), the action of running water and the action of wind.

When a stray body collides with Mars at a typical velocity of 10 kilometers per second, it makes a crater whose diameter is from 10 to 20 times greater than that of the body itself. Over the first billion years of the solar system's evolution the rate of cratering was much higher than it is today. The bodies responsible for the early craters were either objects left over from the formation of Mars and its satellites or objects that had formed in the outer part of the asteroid belt and were perturbed into trans-Martian orbits by gravitational interaction with Jupiter. Currently the small bodies that collide with Mars are derived from comets and asteroids.

A fresh impact crater whose diameter is less than 10 kilometers has a bowl-shaped profile, a raised rim and a hummocky surrounding blanket of ejected material. Craters of larger size usually have a small raised mound in their center; craters larger than about 250 kilometers have multiple raised rims. There is a sharp dichotomy in the density of craters larger than 10 kilometers on Mars: the southern hemisphere shows many more than the northern hemisphere. Evidently the densely cratered terrain in the

southern hemisphere is a much older surface than the one seen in the northern hemisphere. Most of the northern hemisphere has been flooded with lava, which has obliterated the craters that must have existed at one time. It is a curious fact that many other bodies in the solar system, such as the earth and the moon, have also developed asymmetrically.

From estimates of the present rate of crater formation on Mars we can infer that most of the large craters in the southern hemisphere were made early in the history of the planet, probably during the first billion years. During the early intense-bombardment period Mars occasionally collided with very large meteoroids. The impacts created the large basin areas found in or near the southern hemisphere, such as Hellas (45 degrees south, 290 degrees west), Argyre (50 degrees south, 45 degrees west) and Isidis (15 degrees north, 270 degrees west). The largest, Hellas, is about 2,000 kilometers across and about four kilometers deep.

From the state of preservation of craters in various regions on Mars it is possible to infer the relative rates of crater obliteration and degradation over the lifetime of the planet. The large craters in the southern hemisphere vary widely in the degree to which they have been degraded. Nearly all the craters on the volcanic plains of the northern hemisphere, however, have a fresh appearance. This difference suggests that the rate of crater degradation was much higher at an early epoch of Martian history, when the large southern craters

were being made, than it was at times later than the period when the early volcanic plains were formed. Detailed studies of crater morphology show that near the end of the initial cratering epoch the rate of crater degradation increased greatly. The period of enhanced erosion came close to the time of formation of the oldest volcanic plains.

The melting of rocks in the interior of Mars has given rise to two types of volcanic landforms: the large volcanic plains and conelike structures. Part of the reason that lava flooding took place preferentially in the northern hemisphere may be that much of the northern hemisphere lies several kilometers below the average level of the southern hemisphere. The lava plains bear a striking resemblance to the lava plains (the maria, or "seas") on the moon. Lava flowing across these plains has created numerous wrinkle ridges. The fronts of individual flows are characterized by long, low lobed escarpments. Some areas of the northern volcanic plains are pocked with up to 10 times as many craters as other areas. This marked variation implies that the lava flooding has occurred episodically over a long period of time that extends from the formation of the first volcanic region, more than a billion years ago, up to the present.

Among the most impressive features of the Martian surface are enormous shield volcanoes hundreds of kilometers in diameter and higher than the earth's tallest peaks. The youngest and largest is Olympus Mons [see *illustration on page 85*], which is about 600 kilometers across, or approximately three times larger than the closest terrestrial equivalent, Mauna Loa on Hawaii. The summit of Olympus Mons rises 25 kilometers above the surrounding terrain, or more than two and a half times the height of Mount Everest above sea level. Near the summit of the shield is a complex of calderas: collapse features that were once vents for lava. The flanks of Olympus Mons exhibit a rough texture that has numerous radially elongated lobes and remnant lava channels that were created during the last periods of the construction of the shield. The edge of the shield is marked by an escarpment that is two kilometers high. Presumably the cliff face was formed by erosional processes that operated after the formation period.

There are three other prominent shield volcanoes close to Olympus Mons. All are located on the Tharsis ridge (zero degrees latitude, 115 degrees west) and were formed after the creation of the ridge. A much more degraded shield, Alba Patera, is found to their north. Less prominent shields are also visible in the older, elevated Elysium region (20 degrees north, 210 degrees west). If we assume that the Martian shield volcanoes grew at the same rate as terrestrial volcanoes, about 100 million years would have been required to build each of the largest ones. Shield construction did not take place synchronously, even in the Tharsis region, but was spread out over a much longer time than is required to build an individual shield volcano.

A number of smaller volcanic domes, about 100 kilometers across, can also be identified in the Mariner pictures. The domes have much steeper flanks than the shield volcanoes and were probably formed by a more viscous lava. On a still smaller scale there are volcanic craters with scalloped outlines and smooth, well-developed rims.

It seems that most of the shield volcanoes were formed after and partly as a result of tectonic activity that created large elevated areas. Coincident with the formation of these high areas, numerous elongated cracks or fractures in the ground resulted from the stretching of the surface. The most prominent of the tectonic events was centered in the region of Labyrinthus Noctis (five degrees south, 100 degrees west), where it raised the surface by about 10 kilometers. This elevated area, extending over thousands of kilometers, evolved asymmetrically with the emergence of a rather sharp ridge in the Tharsis region, where the ground was stressed more severely than elsewhere. The excessive stress may account for the development of the three prominent shield volcanoes in the Tharsis region. The Tharsis updoming, which may have been engendered by a rising current in the molten portion of the interior, caused a serious fracturing of the surface as far away as the Elysium region. Places where this event or similar events gave rise to numerous fractures are called fossae on a geological map of Mars. Following the development of the elevated plateau, lava plains were laid down and shield volcanoes arose within the rise and close to it. At an earlier epoch in Martian history, which came after the first formation of the volcanic plains, the Elysium rise was created by a similar process, and this too was later accompanied by a set of volcanic events.

Martian tectonism differs in a very important aspect from tectonism on the earth. The earth's crust is divided into a series of plates that are slowly moving with respect to one another and constantly rearranging the earth's geography. On Mars there is no sign of horizontal crustal motions. The lack of plate motion on Mars may explain why the planet's shield volcanoes are so gigantic. On the earth the plate on which a volcanic shield is being built is in motion with respect to the deeper portions of the interior that are supplying the material for it. As a result long strings of shield volcanoes are built, such as the Hawaiian-Emperor chain that stretches from the Aleutians to the Hawaiian Islands.

One of the most remarkable results of the Mariner 9 mission was the discovery of sinuous channels that appear to have been formed by running water. The largest are up to 1,500 kilometers long and as much as 200 kilometers wide. These spectacular features are totally unrelated to the Martian canals so often reported by early observers. Unlike the canals, which supposedly radiated all over the planet, the large channels are found preferentially in the equatorial regions, where the annual temperatures are the highest.

Evidence that the channels were formed by fluvial erosion is given by the presence of immature tributary systems,

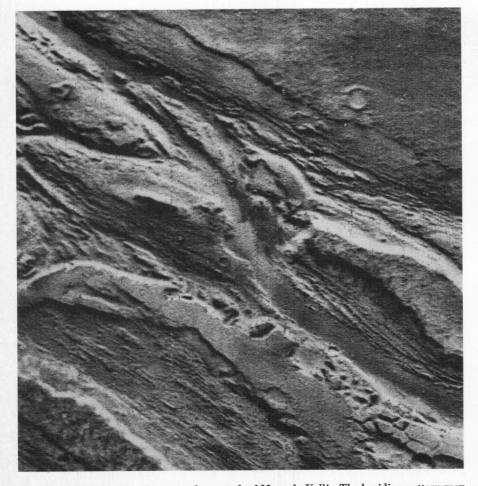

BRAIDED TERRAIN is shown in photograph of Mangala Vallis. The braiding pattern may have resulted from the deposit of silt during the waning stages of the channel's formation. A similar pattern is found in terrestrial riverbeds that are formed during sudden floods.

teardrop-shaped islands and bars, and braided patterns. The braided patterns may have resulted from the deposition of debris carried by the fluid during the waning stages of channel formation. A given channel usually exhibits some of these characteristics but not all of them. Almost without exception the downhill direction of the channel coincides with the direction that would be expected from a fluvial origin, as is implied by the channel's morphological features, such as the geometry of its tributaries.

The Martian channels are quite dissimilar in appearance from lava channels found on the moon and the earth. Therefore the Martian channels were almost certainly not made by flowing lava; a less viscous fluid was required. The most likely candidate is liquid water. Under present conditions, however, the amount of water vapor in the atmosphere of Mars is so small that surface water would quickly evaporate. Evidently the channels were carved at an earlier time, when the Martian climate was warmer and wetter.

In some places there is an obvious source of water for the large channels. For example, the channels in the belt from 10 to 50 degrees west near the equator start at a type of landform called chaotic terrain. Chaotic terrain is a depressed region characterized by a disorderly array of broken slabs of rock. It is thought to have been formed by the withdrawal of subsurface material and the subsequent decay of ground ice. The melting of the ground ice could have supplied the water that cut these channels. For many other channels, however, there is no obvious source of water. In the case of channels with tributaries the water evidently originated over a large area.

The large Martian channels are most similar to features on the earth that have been formed by the sudden flooding of an area, such as the "channeled scablands" in the state of Washington that were carved when a natural dam holding the water in a large ice-age lake gave way. The immature tributary system of some of the channels is also consistent with a short formation time. In some instances there is a suggestion that several flooding events rather than a single such event may have been involved in the cutting of individual channels.

Gullies that are some tens of kilometers long are a second category of features that may have been created by running water. Widely distributed across the surface of Mars, they resemble drainage channels found on surfaces of moderate slope in deserts on the earth. Since

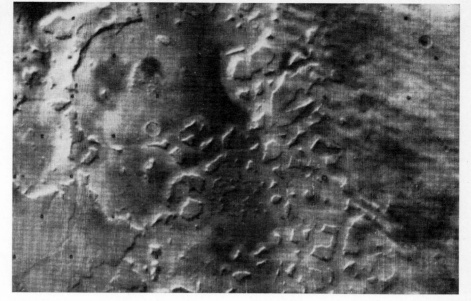

FRETTED HUMMOCKY TERRAIN is a lowland region dotted with a series of isolated mesas. It is bounded by cliffs that have an intricate geometry. This landform may have resulted when ice below the surface decayed, triggering landslides that caused cliffs to retreat.

they are often located on the sides of impact craters, they cannot readily be attributed to flowing lava. Thus the Martian gullies may have been carved by the runoff of rainfall, which would imply a climatic variation even more drastic than the one postulated for the creation of the large channels.

"Fretted terrain" is another landform whose development may have been influenced by sapping, or the disappearance of subsurface ice. The term fretted is used to describe a flat lowland that is bordered by steep cliffs with an intricate geometry. Within the lowland immediately adjacent to the cliffs are numerous small elevated plateaus. A few valleys dissect the highland near its border with the lowland. It is thought that the lowland has been generated at the expense of the highland through the exposure of ground ice at the cliff faces. As the ground ice decays it undermines the top parts of the cliff, which subsequently slump to the bottom. The recession of the cliff faces and their continued decay leads to the creation of the lowland regions. Stranded plateaus in the lowlands represent places where disintegration of highland material is incomplete.

One of the most spectacular features on Mars is a region of enormous canyons just south of the equator between longitudes 45 degrees west and 90 degrees west. The entire canyon system, called Valles Marineris, is about 2,700 kilometers long and in places is 500 kilometers wide. Individual canyons are up to several hundred kilometers long, 200 kilometers wide and six kilometers deep.

The canyon system is thought to be the complex result of fracturing, magma withdrawal, sapping and fluvial erosion. The tectonic event that caused an uplift in the area centered around Labyrinthus Noctis gave rise to a system of fractures that parallel the long axis of the canyon system. Subsurface withdrawal, perhaps the withdrawal of magma needed to build the volcanic shields in the Tharsis region, first created parallel sets of pits along the fractures, which were enlarged and joined together by further withdrawal and wall recession due to sapping and subsequent avalanches. Lines of pits are currently found on the highland margin of the canyons and may represent canyons at an early stage of formation [see illustration on page 86]. The avalanches left narrow, vertical notches near the top of the canyon walls and a blanket of jagged material at the base of the walls. The withdrawal, sapping and avalanche processes created little canyons, which continued to grow by these same processes. The plateaus running down the middle of some of the larger canyons may represent places where two small canyons merged. Running water produced by the decay of ground ice may have played a role in the development of the branched pattern found in the highland area next to the large canyons and in the production of the steep ridges that form the sides of the intra-canyon plateaus.

Wind-driven particles have also played an important part in the shaping of the Martian surface. The observed occurrence of both local and global dust storms offers direct evidence that wind-blown dust is an important force on Mars. When Martian wind velocities exceed 100 miles per hour, small dust particles are set in motion. Particles larger than a thousandth of an inch saltate, that is, travel a small distance through the atmosphere, hit the surface and bounce back into the atmosphere. Smaller particles settle out of the atmosphere so slowly that they tend to stay in suspension and to travel great distances before they sink to the ground.

Viewed through a telescope, Mars consists of continent-size light and dark areas, which show seasonal and longer-term changes in their contrast and in the location of their boundaries. The seasonal changes suggested to some early observers that they were viewing the growth and decay of Martian organisms. Ten years ago, however, Carl Sagan of Cornell and I suggested that the surface changes were due to windblown dust. Our hypothesis has since been supported by *Mariner* 9 pictures, which show that many areas consist of dark and light splotches and streaks that are often aligned in a common direction [*see illustration on page 80*]. Winds blowing in

a fixed direction are responsible for this alignment. Furthermore, small dark features that were not present in an early set of photographs appear in photographs made several weeks later.

The global dust storms play an important part in the seasonal changes of the light and dark markings. As the dust storm decays a uniform blanket of small bright particles settles over all areas. Subsequently winds, which are strong in particular regions because of the region's topography or its closeness to the polar cap, stir some of the small particles back into suspension on a local scale and thereby cause a darkening of the area from which they have been removed. Darkening continues until the next global dust storm.

Streaks and splotches are visible in the vicinity of many craters. Experimental tests carried out in a wind tunnel by Ronald Greeley of the University of Santa Clara, James D. Iversen of Iowa State University and me show that these patterns are created when a dust-filled wind flows near the rim of a crater. As the wind sweeps past the sides of the crater a vortex is generated that preferentially removes particles from the sides and also downstream from the crater along two

zones displaced to each side of the crater's center line. Just behind the center of the downstream rim and on the upstream interior of the crater, however, the air is very still, and particles preferentially accumulate in those zones.

Over much longer periods the action of wind-driven particles has caused both erosion and deposition of material on Mars that have influenced much more than the surface layers of the ground. Since the Martian winds reach higher velocities than terrestrial winds, sandblasting is more effective on Mars than it is on the earth. Evidence for the erosional effects of saltating sand is provided by the parallel grooves and flutings found on the surface of many Martian areas, by the existence of streamlined hills and by the hollows and pits observed within the polar region.

Winds may have put into suspension some of the material that was removed from the fretted terrain and canyon lands of Mars and then carried this material to the polar regions, where it formed a thick sedimentary blanket. The polar deposits are composed of a mixture of small dust particles and ices. There are two main sedimentary deposits in both polar areas that cover the older terrain in those regions. The older blanket is an unlayered deposit that covers pitted plains. The younger one lies closer to both poles and is the geological unit on which the permanent polar caps lie. Because of its layered texture it has been termed laminated terrain. In pictures taken with the wide-angle, low-resolution camera on *Mariner* 9 the laminated terrain seems to consist of a series of stacked plates, with successively smaller plates resting on top of larger ones. In high-resolution photographs the walls of individual plates are seen to consist of individual layers, each about 30 meters thick. The layered structure suggests that there have been periodic variations in the processes that laid down the younger blanket. At present both polar deposits are being eroded. The material that has been removed has preferentially settled at middle-latitude regions surrounding the polar blankets, where it forms a thin veneer on the surface and partly fills the interior of small craters.

From several independent kinds of evidence it seems inescapable that the climate of Mars was once very different from what it is today. One mechanism for climatic change on Mars is based on the great sensitivity of atmospheric carbon dioxide to changes in the polar temperatures. When the polar temperature

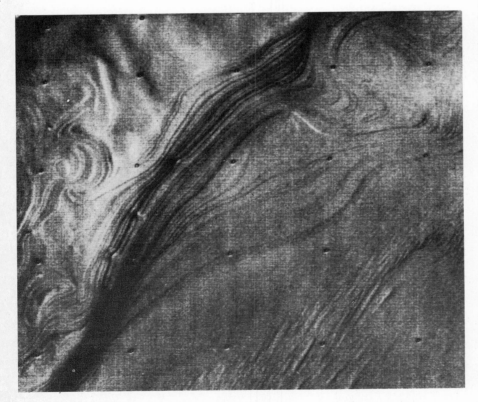

**LAYERED TERRAIN in the south-polar region of Mars appears as the banded structure running diagonally across this photograph. The layered deposits have been exposed on an inclined surface that has a staircase topography. Erosion by the Martian wind has smoothed the surface until it assumed this gracefully sculptured appearance and may also have carved the grooves on the adjacent flat surface near the bottom of the photograph. The layered structure implies that there was a periodic variation in deposition processes that formed it.**

is low, the amount of carbon dioxide in the atmosphere drops. When the temperature is high, the carbon dioxide content may increase greatly if there is a large enough reservoir of carbon dioxide in the regolith to supply the equilibrium amount of atmospheric carbon dioxide. Peter J. Gierasch, Owen B. Toon and Sagan have shown that an initial increase in polar temperatures increases the total mass of the atmosphere, which then carries more heat to the pole and warms it further. In some cases the feedback process gives rise to a runaway situation leading to a much higher atmospheric pressure, a pressure comparable to the atmospheric pressure on the earth.

One can think of several factors that could alter the polar temperatures and therefore lead to very large changes in the carbon dioxide content of the Martian atmosphere. First there is the possibility that the luminosity of the sun has changed at intervals of millions to hundreds of millions of years. Any such change in the output of the sun would naturally cause synchronous changes in the climate of all the planets, including the earth. Second, the reflectivity of the Martian polar regions may change because of variations in the frequency of major dust storms, which deposit material in those regions. Third, as William R. Ward of Harvard University has shown, Mars's axis of rotation, unlike the earth's, changes its inclination toward or away from the sun by as much as 20 degrees in cycles of 100,000 and a million years. These changes would seem too fast, however, to explain many interesting features, including the layers of the laminated terrain.

Whatever the cause, an increase in the content of carbon dioxide in the Martian atmosphere will tend to force water to go through the liquid state instead of vaporizing directly from the frozen state. On the other hand, an increase in atmospheric carbon dioxide may actually lower the temperature in the equatorial regions by diminishing the equator-to-pole temperature gradient. That would militate against the appearance of liquid water. Moreover, carbon dioxide by itself can do little more than it currently does to warm the Martian surface by the "greenhouse effect," where the atmosphere absorbs some of the heat the surface is emitting and radiates it back to the surface. Any process that increases the atmosphere's carbon dioxide content, however, would simultaneously raise its water-vapor content, which would effectively warm the Martian surface by the greenhouse effect. Thus plausible mechanisms seem to be available for creating the climatic conditions needed for running water to be stable in the equatorial regions of the Martian surface.

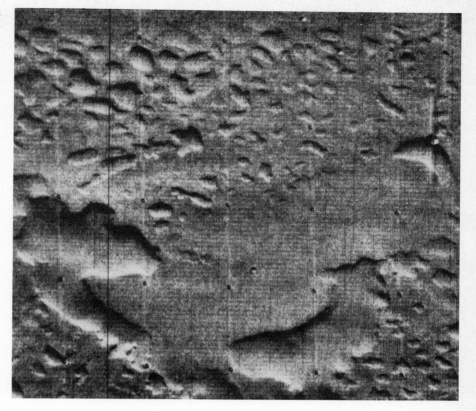

ETCHED PLAINS in the south-polar region are older than the layered terrain. Furthermore, unlike the layered terrain, the depositional blanket shows no obvious layering where it has been exposed by erosion from the wind eating away the walls of the pits and hollows.

A final factor that may be important for climatic change on Mars is an alteration of the atmosphere from its present oxidized state to a more reduced state. I have already suggested that when the atmosphere of Mars was being outgassed from the interior of the planet some of the carbon atoms were present in the form of methane rather than carbon dioxide. It is conceivable that ammonia was also present. Since methane and ammonia have lower condensation temperatures than carbon dioxide does, such a reducing atmosphere could easily reach high pressures. Ammonia acting together with water vapor would very effectively warm the surface by the greenhouse effect. It is tempting to speculate that an early reducing atmosphere was a by-product of the formation of the oldest volcanic plains and that it was responsible for the era of enhanced crater degradation. Conditions may have been favorable at that time for the formation of the fluvial channels and gullies. As the outgassing slowed down, the molecular-hydrogen content of the atmosphere was greatly diminished, and the Martian atmosphere evolved to its present oxidized state.

Such a reducing atmosphere is thought to have been needed in the early history of the earth to account for the emergence of the first forms of life on our planet. Could life have also been generated on Mars when the planet had the hypothetical reducing atmosphere? Could that life have adapted to the more stringent environmental conditions of the present? Although there is no liquid water on Mars today, could a hypothetical Martian organism make use of the water that is bound in the surface sand grains? We do not know the answers to these questions. The Viking lander that is scheduled to descend gently to the Martian surface next July will conduct several experiments designed to test for the presence of life. Because we do not know the precise characteristics of Martian life, if it exists, our hopes for obtaining a definitive answer from the Viking mission should not be too high. At the very least, however, we shall gain important new knowledge about the physical environment on Mars. And by learning more about the factors that influence climatic change on Mars we may be able to better understand the factors that cause climatic change—past, present and future—on our own planet.

# 9

## JUPITER

# Jupiter

JOHN H. WOLFE

*More massive than all the other planets put together, it consists largely of hydrogen and helium. Below its turbulent atmosphere the hydrogen forms two liquid layers, one molecular and one metallic*

The most conspicuous property of Jupiter is its size: it is not merely the largest planet but has almost two and a half times the mass of all the other planets put together. In the evolving solar system it acquired most of the matter that did not go into the formation of the sun. Indeed, it does not introduce gross error to regard the solar system as a two-body association, neglecting everything but the sun and Jupiter.

The characteristics of an object that makes up so much of the solar system are of obvious interest. The elemental composition of Jupiter is known to be similar to that of the sun, but the structure of the planet resembles neither that of a star nor that of the terrestrial planets, such as the earth. Jupiter is essentially a liquid body, with at most a small solid core. Above the surface of the liquid is a thick atmosphere, whose storms and patterns of circulation appear to be similar in principle to those on the earth but whose scale is so much larger that comparison seems extravagant. Jupiter has an elaborate magnetic field and a complex system of radiation belts, and it is a prodigious emitter of radio waves.

Finally, Jupiter has 13 moons, more than any other planet. The discovery of the largest four of these satellites was a signal event in the history of astronomy. In 1610 Galileo Galilei trained his newly constructed telescope on Jupiter and was astonished to see these bright moons, which were subsequently named Io, Europa, Ganymede and Callisto. Galileo's observation provided the first evidence that a planet other than the earth could have satellites, and it completely discredited non-Copernican theories of celestial mechanics.

Since the time of Galileo astronomers have studied Jupiter intensively in the visible, infrared and radio-frequency regions of the spectrum. More recently two spacecraft in the U.S. Pioneer series have passed near the planet, sending back photographs and other kinds of information. *Pioneer 10* was launched in March, 1972, and made its closest approach to Jupiter in late November and early December, 1973; *Pioneer 11* was launched in April, 1973, and flew past the planet approximately a year after *Pioneer 10.*

A first reason for supposing that the composition of Jupiter should resemble that of the sun is a cosmogonic one: If the sun and the planets all formed from a single cloud of gas and dust, then initially each probably received an approximately equal share of all the materials in the cloud. The composition of the smaller planets has been drastically altered by the loss of the lighter elements, but Jupiter is so massive that it could have retained virtually all of even the lightest element, hydrogen. The time required for any significant quantity of hydrogen to escape from Jupiter's gravitational influence is many orders of magnitude greater than the age of the solar system. Jupiter, like the sun, should have approximately the same elemental composition today that it had when it formed some 4.6 billion years ago.

This argument is supported by measurements of the density of Jupiter. Although Jupiter has more than 1,000 times the volume of the earth, it is only about 318 times as massive, so that its density (1.33 grams per cubic centimeter) is less than a fourth that of the earth. Jupiter must therefore be made up largely of light elements, and the only plausible candidates are hydrogen and helium. Of fundamental importance in any model of its structure is the ratio of hydrogen to helium. That ratio has been measured by several techniques by instruments on *Pioneer 10* and *Pioneer 11.* Although the measurements are not in strict agreement, they are all consistent with the hypothesis that the ratio on Jupiter is the same as it is in the sun. The solar ratio is roughly one atom of helium for 10 molecules of hydrogen.

Models of the interior of Jupiter must take into account several lines of evidence in addition to elemental composition. For example, Jupiter has been found to radiate about twice as much heat as it receives from the sun, and so it must therefore have an internal source of heat. The Jovian magnetic field, an order of magnitude more powerful than the earth's, must also be explained; present theories of how planetary magnetic fields are generated require that the interior of the planet contain an electrically conductive fluid. Finally, a model of the planet's structure must be consistent with the observed shape and intensity of the planet's gravitational field.

If Jupiter were not rotating, it would assume the form of a perfect sphere (neglecting perturbations caused by satellites and by the sun) and would act as a point mass. Its gravitational field would exhibit spherical symmetry: at a

NORTH POLE OF JUPITER (*opposite page*), which cannot be seen from the earth, was photographed for the first time by the *Pioneer 11* spacecraft. The pole is near the line of the terminator, the boundary between day and night at the top of the picture. The dark blue-gray areas at the top and sides may be "blue sky" caused by Rayleigh scattering of sunlight by the atmosphere. The north-polar region lacks definitive banding, and its mottled appearance suggests that there are rising convection cells. Photograph was made from a distance of 1.3 million kilometers (800,000 miles) and at a latitude of 50 degrees above the equator.

BANDS OF JUPITER consist of light "zones" that range in color from pure white to pale yellow and dark "belts" of various shades of reddish brown. The zones and belts near the equator are permanent bands that vary slowly in width and color. They occasionally split into sub-bands. The most prominent marking on Jupiter is the Great Red Spot, which has been observed for more than 300 years since its discovery by Giovanni Domenico Cassini. This photograph was made by *Pioneer 10* from three million kilometers (1.8 million miles).

"LITTLE RED SPOT" in the north tropical zone of Jupiter was photographed in 1973 by *Pioneer 10*. Its structure is similar to that of the Great Red Spot in the south tropical zone. The similarity takes away some of the uniqueness of the Great Red Spot and suggests that both phenomena may be long-lived cyclonic features akin to a hurricane. The Little Red Spot was observed from the earth for 18 months before it was photographed by *Pioneer 10*. It disappeared before *Pioneer 11* reached Jupiter and thus had a lifetime of about two years.

given distance from the planet the intensity of the field would be the same when it was measured above the equator or above the pole or at any other point. In actuality Jupiter is rotating rapidly. The rotation distorts the form of the planet and thereby alters the contours of the gravitational field.

Jupiter's period of rotation can be determined simply by measuring the interval between successive transits across the central meridian of some long-lived feature in its atmosphere. Such measurements, however, give different results for different latitudes. Markings within 10 degrees of the equator appear to rotate with an average period of nine hours 50 minutes 30 seconds. For features at higher latitudes the apparent period of rotation is slightly longer; the adopted value is nine hours 55 minutes 41 seconds. The discrepancy is probably caused by systematic variations in the vast Jovian winds.

A more reliable and perhaps more meaningful measurement of the rotational period can be obtained through observations of the planet's radio-frequency emissions. Some of these emissions are polarized and directionally oriented by the Jovian magnetic field. Since the magnetic dipole axis is inclined by about 11 degrees to the rotational axis, the signals received on the earth vary in intensity and in direction of polarization in synchrony with Jupiter's rotation. The period derived from radio measurements is nine hours 55 minutes 30 seconds, in good agreement with the value measured optically for higher latitudes. What is actually measured by this method is the rotation of the magnetic field. Since the currents that generate the field are probably rooted deep within the planet, radio-frequency observations probably represent more accurately than other methods the rotation of Jupiter as a whole.

Jupiter's low density and rapid rotation give rise to a pronounced equatorial bulge. The best determinations of the planet's shape, measured to the cloud tops, give an equatorial radius of 71,400 kilometers and a polar radius of 67,000 kilometers. Thus Jupiter is flattened at each pole by 4,400 kilometers, about two-thirds the radius of the earth. (For astronomical measurements in the vicinity of Jupiter a standard radius of 71,372 kilometers has been adopted.)

The degree and the nature of Jupiter's deviations from spherical form are determined by the density distribution in the interior of the planet. That distribution must also be reflected in the planet's external gravitational field, and the study of the field has proved to be a powerful tool for deducing the structure of the planet.

The detailed form of the Jovian gravitational field has been examined by tracking the Pioneer spacecraft as they flew past the planet. Their trajectories could be calculated on the assumption that Jupiter is a point mass, and deviations from the predicted trajectories could be detected with great precision. The gravitational analysis made possible by the Pioneer tracking experiments established that the solid body below Jupiter's atmosphere is in fact liquid.

From the Pioneer observations a model of Jupiter's interior has been devised by John D. Anderson of the Jet Propulsion Laboratory of the California Institute of Technology and William B. Hubbard of the University of Arizona [*see illustration below*]. The model is consistent with what is known of Jupiter's gravitational and magnetic fields and with extrapolations of laboratory studies of the behavior of hydrogen at high temperature and pressure to even higher values.

The model allows for a small rocky core at the center of the planet, where the temperature is thought to be about 30,000 degrees Kelvin. The core would be composed mainly of iron and silicates, the materials that make up most of the earth's bulk. Such a core is expected for cosmogonic reasons: if Jupiter's composition is similar to the sun's, then the planet should contain a small proportion of those elements. Since they are relatively dense, they would aggregate at its center. The core cannot be detected through gravitational studies, however, so that its existence cannot be proved.

Above the hypothetical core is a thick stratum in which hydrogen is by far the most abundant element; this stratum makes up almost all the mass and volume of the planet. The hydrogen is separated into two layers; in both it is liquid, but it is in different physical states. The inner layer extends from the core to a distance of approximately 46,000 kilometers from the center, where the pressure is estimated to be about three million earth atmospheres and the temperature near 11,000 degrees K. In this layer the hydrogen is in the liquid me-

tallic state, a form of the element that has not yet been observed in the laboratory because it exists only at extremely high pressures. In the liquid metallic state hydrogen molecules are dissociated into atoms and the fluid is an electrical conductor.

The outer layer extends to about 70,000 kilometers and consists mainly of liquid hydrogen in its molecular form. Above the layer of molecular hydrogen, and extending another 1,000 kilometers to the cloud tops, is the atmosphere.

If this model of the structure of the interior is accurate, the excess heat radiated by Jupiter is simply a remnant of the heat generated when the planet coalesced from the solar nebula. Earlier models invoked radioactivity or the heat liberated by the contraction under gravitational forces of a largely gaseous planet. Since a liquid is virtually incompressible, however, Jupiter cannot be radiating heat because it is contracting; on the contrary, it is contracting because it is slowly cooling. This explanation of the excess radiation requires that the planet's primordial thermal energy, which is confined mainly to the interi-

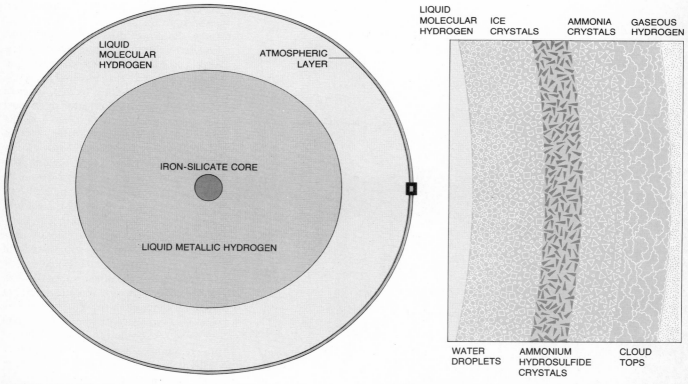

**MODEL OF JUPITER'S INTERNAL STRUCTURE,** shown here in cross section, is based on the assumption that liquid hydrogen makes up the bulk of the planet's interior except for a possible small, iron-silicate core. The existence of such a core is based on the argument that the abundance of the elements in Jupiter must be similar to that in the sun. The model predicts a temperature of 30,000 degrees Kelvin in the core region. A thick shell of liquid metallic hydrogen surrounds the core. Metallic hydrogen is an electrical conductor, and electric currents in the shell may be the source of Jupiter's magnetic field. At approximately 46,000 kilometers from the center of the planet there is a transition from liquid metallic hydrogen to liquid hydrogen in its molecular form. The pressure in the transition region is about three million atmospheres, and the temperature is estimated to be about 11,000 degrees K. The shell of liquid molecular hydrogen is about 24,000 kilometers thick. Above the liquid hydrogen lies Jupiter's gaseous atmosphere, which is about 1,000 kilometers thick. An enlarged detail of the transition from liquid hydrogen to gaseous hydrogen is given at the right.

or, be steadily conveyed to the surface. Thus the liquid-planet model predicts that most of Jupiter's mass is stirred by large-scale convection currents, although convection might be inhibited by a gradient of helium concentration.

Convection currents also provide a probable mechanism for the generation of the Jovian magnetic field. Convection currents within the layer of liquid metallic hydrogen, deflected by the Coriolis force, could set up loops of electric current, which would give rise to a magnetic field. A similar process in the earth's core is thought to generate the terrestrial magnetic field, but the theory of such planetary dynamos is not well established.

Neither the surface nor the interior of Jupiter is accessible to us today; when we look at the planet, we see only the top of the atmosphere. The visible disk of the planet is covered with the familiar sequence of alternating light and dark bands, all oriented parallel to the equator. By convention the light features are called zones and the dark ones belts. The zones are generally pure white to pale yellow and the belts are various shades of reddish brown. The five zones and four belts in the central portion of the disk are permanent features, although they vary slowly in width, color and intensity. At somewhat higher latitudes the bands are less permanent, and at latitudes beyond about 50 degrees the banding pattern disappears entirely, to be replaced by a less orderly structure in the polar regions.

The surface is marked not only by bands but also by a variety of plumes, streaks, swirls, loops, spots and irregular patches, all varying in color from white to reddish brown to red. The most prominent of these markings is the Great Red Spot, an immense oval feature contained largely within the south tropical zone but protruding somewhat into the south equatorial belt [*see illustration on cover of this book*]. The spot has persisted for at least 300 years, since it was discovered by Giovanni Domenico Cassini in 1665. Its width is fairly constant at about 14,000 kilometers, but its length varies from 30,000 to 40,000 kilometers over a period of a few years. The intensity of its coloration is also variable: during the Pioneer encounters of 1973 and 1974 it was prominent, whereas six to seven years earlier it had appeared to be quite faint.

The Pioneer missions revealed that in the atmosphere, as in the interior, convection plays a crucial role. In passing around the planet, the spacecraft measured atmospheric temperatures on the dark side of Jupiter, measurements that cannot be made from the earth. (Since Jupiter's orbit is far beyond the earth's, we observe the planet almost as if we were on the sun and we see little but the illuminated portion of the disk.) It was found that the day side and night side are at the same temperature, suggesting the enormous heat capacity of the Jovian atmosphere and confirming the importance of an internal source of heat. It was also determined that the light zones are colder, and therefore higher in the atmosphere, than the dark belts. The temperature measurements provided compelling evidence that the zones are regions of rising gas and the belts are regions of descending gas [*see illustration on this page*].

Even in the polar regions (which also cannot be seen clearly from the earth and which were photographed for the first time by *Pioneer 11*) the clouds are mottled, suggesting that they are divided into numerous convection cells [*see illustration on page 94*]. At lower latitudes the powerful Coriolis forces generated by Jupiter's rapid rotation convert vertical convection currents into horizontal bands that girdle the planet. As the gas rises within a zone, it tends to move toward the equator or toward the pole in order to descend into the adjacent belt. This north-south motion is deflected by the Coriolis force to produce an east-to-west circulation. The same mechanism can be observed in the earth's atmosphere in the development of the trade winds, but the Coriolis force on Jupiter is much stronger. Atmospheric features near the equator have been observed moving with speeds of several hundred kilometers per hour.

Near the center of the zones the flow of gases in opposite directions creates regions of wind shear. The highest velocities (with respect to the interior of the planet) are found in jet streams at the boundaries between zones and belts. The high wind speeds must lead to turbulence, and they are almost certainly responsible for the eddies, loops and swirls visible in the atmosphere. They may also initiate or sustain spots, which are generally somewhat larger than the other features, but whether they can account for the Great Red Spot remains conjectural.

Before the *Pioneer 10* encounter with

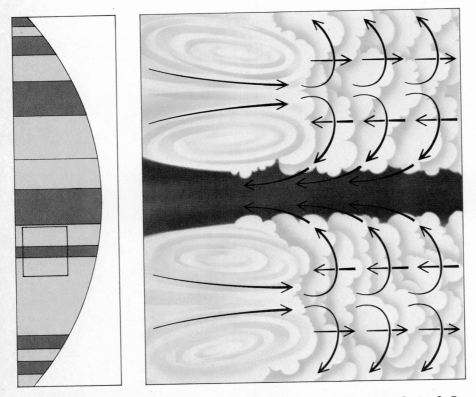

**ATMOSPHERIC CIRCULATION** near the equatorial region of Jupiter is depicted. Gas warmed by the planet's internal heat rises and cools in the upper atmosphere, forming clouds consisting of ammonia crystals suspended in gaseous hydrogen. The clouds form the light zones on Jupiter, which are higher and colder than the dark bands. At the top of the clouds the cooled gas on one side of the light zone tends to descend toward the equator, whereas the gas on the other side of the zone tends to descend toward the pole. The Coriolis forces produced by Jupiter's rapid rotation deflect the north-south motion to the east and west. A similar mechanism in the earth's atmosphere is responsible for the trade winds.

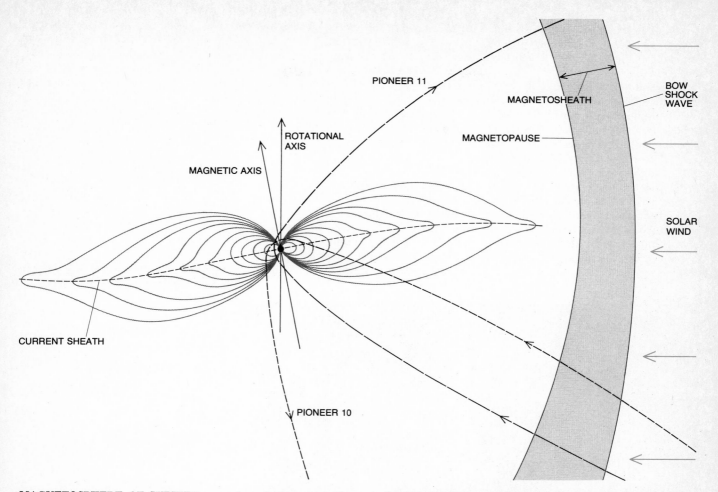

MAGNETOSPHERE OF JUPITER expands and contracts with changes in the impinging pressure of the solar wind. The boundary where there is an equilibrium between the pressure of the magnetic field and the pressure of the solar wind is called the magnetopause. A standing bow shock wave is formed in front of the magnetopause. On their journey past Jupiter, *Pioneer 10* and *Pioneer 11* reported the magnetopause to be as far out as 100 Jupiter radii from the planet and as close in as 50 radii. Jupiter's magnetic field is dipolar, and the axis of the dipole field is inclined 10.8 degrees with respect to the rotational axis. In addition the axis of the dipole field is displaced about 7,000 kilometers from the center of the planet. Measurements by *Pioneer 11* revealed that the strength of Jupiter's magnetic field at the cloud tops ranges from three to 14 gauss. The magnetic field becomes more complicated closer to the planet, displaying quadrupole and octopole moments that presumably are the result of complex circulation patterns in the planet's interior.

Jupiter the Great Red Spot was commonly attributed to a vortex phenomenon called a Taylor column, which explained the spot as a kind of standing wave formed over a mountain or a depression on the surface of the planet. One weakness of this hypothesis is that during the past century the spot has wandered in longitude a distance equal to several circuits of the planet. Moreover, since we now believe the bulk of Jupiter is liquid and could have no mountains or depressions, the Taylor-column hypothesis appears unlikely.

Today it seems more plausible that the Great Red Spot is a cyclonic disturbance somewhat similar to a hurricane. On the earth hurricanes maintain their strength as long as they remain in the tropical ocean; in most cases they deteriorate only when they move over land or colder water. Jupiter has no land, and the Great Red Spot is appar-

ently confined to a narrow range of latitudes, the south tropical zone, which might be considered Jupiter's "hurricane belt." If the spot is a storm, its longevity is somewhat puzzling, but it is possible that the lifetime of any cloud feature is merely a function of its size.

The interpretation of the Great Red Spot as an immense storm is supported by the recent discovery that it is not unique. In 1972 a much smaller spot appeared in the northern hemisphere, and 18 months later it was shown by *Pioneer 10* to be similar in shape and color to the Great Red Spot [*see bottom illustration on page 96*]. A year later, when *Pioneer 11* flew past the planet, the small spot had disappeared, suggesting that its lifetime was about two years.

The discovery that Jupiter's zones and belts consist of gases at different altitudes and temperatures illuminates

their physical structure, but it cannot account for their dramatic coloration. The colors must be explained in terms of the chemistry of the atmosphere, which is imperfectly understood. Five substances in Jupiter's atmosphere have been identified spectroscopically: hydrogen, helium, ammonia, methane and water. The presence of one more, hydrogen sulfide, has been inferred. All these gases are colorless, and so other materials must be present to create the observed patterns. Among the molecules proposed as coloring material are ammonium sulfide and ammonium hydrosulfide, free radicals, various organic compounds and complex inorganic polymers. Many such substances could be formed in the upper atmosphere, but major constituents of the clouds not only must be formed but also must be formed rapidly. Under the influence of strong convection currents, molecules in the Jovian clouds could quickly be carried downward to regions of high temperature, where they would dissociate. If a substance is to be abundant at the cloud tops, it must be regenerated.

The cloud tops in the zones, the highest and coldest features of the visible atmosphere, are probably ammonia crystals. At the temperature of the cloud tops ammonia would be frozen, which could account for the whiteness of the zones. At the somewhat lower altitude of the belts the temperature is above the melting point of ammonia. It is at this level that the various colored compounds would be found. At the next lower stratum water may be present, first as ice crystals, then, still lower, as droplets.

That Jupiter has a magnetic field was first surmised in the 1950's, when radio-frequency emissions from the planet were discovered. The emissions are confined to two relatively broad regions of the spectrum, that of wavelengths measured in tenths of meters (decimetric emissions) and that of wavelengths measured in tens of meters (decametric emissions). A major component of the decimetric emission consists of thermal radiation, which has a continuous spectrum and random polarization; it is constant with time and is emitted by the entire disk. A significant contribution to the decimetric emission, however, is made by a nonthermal mechanism and is dependent on the planet's magnetic field. It consists of synchrotron radiation emitted by electrons moving with speeds near the speed of light in the magnetic field. The relativistic electrons follow helical paths along the magnetic lines of force, radiating away part of their energy as they travel between the magnetic poles.

The decametric radiation is intermittent and its source is not known. It may be generated by electrical discharges in the atmosphere and the ionosphere. Its intensity is modulated by the innermost of the Galilean satellites, Io. Both the decimetric synchrotron emissions and the sporadic decametric radiation are associated with the rotation of Jupiter's magnetic field.

Except near the planet, the major component of Jupiter's magnetic field is dipolar, like the earth's. The direction of the field, however, is opposite to that of the earth's, so that a terrestrial compass taken to Jupiter would point south. The axis of the dipole field is inclined with respect to the rotational axis by about 10.8 degrees, and the center of the axis is displaced from the center of the planet by about a tenth of Jupiter's radius, mainly along the equator. At a distance of three Jupiter radii the strength of the field is about .16 gauss. Closer to the planet the field is more complex, but measurements made by *Pioneer 11* indicate that at the cloud tops it ranges from about three gauss to slightly more than 14 gauss. (The earth's field, in comparison, varies from .3 gauss to .8 gauss at the surface.)

Within about three Jupiter radii the field can no longer be effectively described as a dipole. *Pioneer 11* measurements have shown that it has quadrupole and octopole moments, that is, there are components of the field with four and eight poles respectively. How these higher-order fields are oriented with respect to the planet and to the dipole field is not yet known. Presumably they reflect some complex pattern of circulation within the liquid metallic hydrogen of the planetary interior. They can only be detected close to the planet because they diminish with distance much more rapidly than the dipole field does. The intensity of the dipole field decreases as the third power of the distance, but the quadrupole field diminishes as the fourth power of the distance and the octopole as the fifth power.

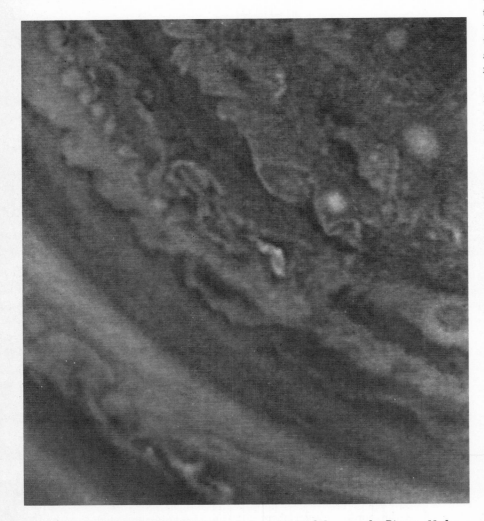

CLOSE-UP OF JUPITER'S SURFACE made from 600,000 kilometers by *Pioneer 11* shows details never seen before. The scallops and swirls between the dark and the light bands are believed to be the result of shearing between adjacent counterflowing jet streams. The white features generally show upwelling of the atmosphere and the dark areas show a downward movement. The white circular spots in the north-polar region at the top are thought to be hurricanelike storms. The Jovian polar storms, like the tropical hurricanes on the earth, could be powered by the latent heat of condensation. A column of water vapor and ammonia vapor warmed by Jupiter's internal heat would begin to rise, creating a low-pressure area. More atmosphere would rush in and be sucked up by the rising column. As the column rises, water vapor would condense, releasing more heat, which would drive the column higher. At the top of the column the remaining heat would be radiated into space.

The magnetic field with its entrained plasma makes up Jupiter's magnetosphere. Outside the ionosphere the planet's environment is all but entirely determined by the magnetosphere and its interactions with the interplanetary solar wind and with the inner satellites. Where the solar wind impinges on the field, equilibrium is established between the pressure of the wind and the internal pressure of the magnetic field and plasma. The boundary between the two regimes is called the magnetopause. Because the solar wind is supersonic with respect to Jupiter (and the other planets) a shock wave forms ahead of the magnetopause [see "Interplanetary Particles and Fields," by James A. Van Allen, page 127].

The geometry of the earth's magnetosphere is essentially similar to that of Jupiter's, and the terrestrial magnetopause forms at a distance of between 70,000 and 80,000 kilometers from the earth. The Jovian magnetosphere is much larger: the distance from the planet to the magnetopause may be as much as 100 times greater. One factor contributing to this difference in scale is that the pressure of the solar wind diminishes as the square of the distance from the sun; since Jupiter is five times as far from the sun as the earth is, the solar wind there is 25 times weaker. Jupiter's magnetic field is also about an order of magnitude stronger than the earth's, and the outer regions of the Jovian magnetosphere are greatly inflated with thermal plasma. If Jupiter's magnetosphere could be seen with the unaided eye, it would appear to be at least twice the apparent size of the moon.

Although Jupiter's magnetosphere is large, it is also quite variable; in particular it is sensitive to relatively minor changes in the pressure of the solar wind. *Pioneer 10* and *Pioneer 11*, while approaching the planet, reported crossing the magnetopause as far as 100 Jupiter radii from the planet and as close as 50 Jupiter radii. For the earth's magnetosphere to shrink or expand by a factor of two is exceedingly rare and would be expected only during the largest of solar magnetic storms. Because of the soft or spongy character of the outer magnetosphere similar magnetic changes on Jupiter appear to be common.

The analysis of Jupiter's electromagnetic environment is further complicated by the presence and motion of satellites within the magnetosphere. Of the 13 known satellites five influence the distribution of charged particles: the four Galilean satellites, and Amalthea, the innermost of the moons. The orbits of the rest are generally beyond the magnetosphere, just as our own moon revolves beyond the range of the earth's magnetosphere. Each of the inner satellites can intercept charged particles and thereby remove them from the population of the radiation belts. As a moon revolves in its orbit it can sweep a corridor clear (while itself acquiring intense radioactivity). Even Amalthea is apparently effective in removing particles from the magnetosphere; its diameter is only about 150 kilometers, but it circles the planet in 12 hours and passes through the core of the radiation belt.

One of the Galilean satellites, Io, has an even more profound effect on the magnetosphere. It not only traps particles but also produces and accelerates them. It has long been known that Io somehow influences the decametric radio emissions detected from the earth. When the satellite is at fixed positions with respect to the earth-Jupiter line, the magnitude of the emissions increases. A possible explanation of this enhancement was proposed when *Pioneer 11* demonstrated that Io has an ionosphere. Because the ionosphere constitutes a conducting fluid, the movement of Io through Jupiter's magnetic field generates a potential across the satellite. Charged particles encountering this potential could be accelerated and thereby induced to emit radio waves. As *Pioneer 11* flew through the magnetic-field lines passing through Io it encountered electrons in sufficient numbers and of sufficient energy to account for the radio emissions. Whether or not the electrons actually produce the observed radiation, however, remains uncertain.

Io's ionosphere is tenuous, but its presence implies that the satellite also has an atmosphere, even if it is one of very low pressure. A partial torus of neutral hydrogen has been observed extending on each side of Io along its orbit. These findings were surprising, since Io is only about as large as the earth's moon, which has not been able to retain an atmosphere. The explanation probably lies in Io's position within the Jovian radiation belts. As gases escape from Io they could be continually replenished with ions captured from the magnetosphere.

Inside Amalthea's orbit, at 2.5 Jupiter radii, the flux of energetic electrons and protons becomes more complicated, just as the magnetic field does. The particle density does not rise to a single maximum as one approaches the magnetic equator or the surface of the planet. Instead the particle flux varies from place to place in a complex way, with concentrations of particles at many points. In the absence of any satellites inside Amalthea's orbit the observed particle distribution can be explained only by irregularities or higher-order moments in the magnetic field.

In the outer magnetosphere, beyond about 20 Jupiter radii, the distribution of charged particles has also turned out to be more complex than was initially supposed. *Pioneer 10*, which passed by Jupiter on an equatorial trajectory, indicated that the flux of energetic electrons was intense near the magnetic equator but elsewhere fell almost to interplanetary values. This observation suggested that Jupiter's magnetic field confines the particles to a narrow sheet near the magnetic equator. *Pioneer 11*, which was to leave the magnetosphere at high latitude, was therefore expected to encounter only low intensities of energetic electrons. Actually the electron flux detected during the outbound voyage of *Pioneer 11* was higher than that observed by either spacecraft at any other time.

High-energy electrons were detected by the Pioneer spacecraft in another surprising place: ahead of the bow shock wave in interplanetary space. Apparently some fraction of the electrons can escape from the outer magnetosphere. When these electrons were discovered, the records of early earth satellites were reexamined and found to contain evidence that the particles travel all the way to the earth. The satellite data showed an enhancement in the background level of cosmic-ray electrons about every 13 months, an anomaly that had not been understood when the data were recorded. We now know that the significance of the interval is that it relates the orbits of the earth and Jupiter: every 13 months the two planets are connected by the spiral lines of the interplanetary magnetic field. The electrons are clearly of Jovian origin. Even the earth cannot escape the influence of the largest planet.

Many of the riddles and ambiguities that remain in our interpretation of Jupiter will no doubt be resolved during the next several years, both by earth-based observations and by spacecraft. For example, a high-resolution television camera to be carried aboard a craft scheduled to fly by Jupiter later in the decade should provide more detailed information on the structure of Jupiter's clouds and on the surface features of the Galilean satellites. More precise knowledge of the structure and composition of the Jovian atmosphere and of the nature of the magnetosphere will probably have to await an atmospheric entry probe and a Jupiter orbiter now planned for the 1980's.

# THE OUTER PLANETS

# The Outer Planets

DONALD M. HUNTEN

*Beyond Jupiter are the remote unexplored planets: Saturn, Uranus, Neptune and Pluto. Saturn has a composition much like Jupiter's; Uranus and Neptune appear to be rockier. Pluto is a small maverick*

The five outer planets of the solar system differ radically from the four inner planets. The inner planets Mercury, Venus, the earth and Mars are all in the same size range and have a high density. Jupiter, Saturn, Uranus and Neptune are remarkable both for their immense size and for their low density. Pluto, the outermost planet, is about the same size as Mercury and is unusual in that its orbit is tilted 17 degrees with respect to the mean central plane of the solar system. No other planet has an orbital tilt greater than seven degrees. Jupiter is taken up in the preceding article. Here I shall discuss what is known about the nature of the four planets beyond Jupiter.

Saturn, the sixth planet from the sun, was the most distant one known to man until the 18th century. It is yellowish in color and brighter than most stars. Uranus, the seventh planet, was discovered in 1781 by William Herschel during a telescopic survey of the sky. He immediately recognized it as a planet and within a few nights he had also observed its motion.

The discovery of Neptune, one of the great triumphs of celestial mechanics, is also a fascinating study in psychology. In 1841 John Couch Adams, who was then an undergraduate at Cambridge, undertook calculations to demonstrate that an unknown planet was perturbing the motion of Uranus. When he communicated his results to astronomers four years later, his results were not taken seriously, and no one looked for a planet at the predicted position. Instead a search of a large area of the sky was undertaken, and delays in interpreting the observations caused the discovery to be missed. Meanwhile in France, Urbain Leverrier had taken up the problem. In 1845 he also published calculations that gave a position for the unknown planet almost identical with the one predicted by Adams. Leverrier could not persuade observers in Paris to interrupt their other work. Eventually he wrote to a young astronomer in Berlin, Johann Gottfried Galle, who found the planet in his first attempt on September 23, 1846. It was less than one degree from the predicted position.

Planetary astronomers today find the story of Uranus' discovery curiously reminiscent of their own experience. Most astronomers are exclusively interested in stars and galaxies and are reluctant to let their telescopes be "wasted" on the planets. It is partly for this reason that the National Aeronautics and Space Administration has financed the building of four major new telescopes, one in Arizona, one in Texas and two in Hawaii.

The discovery of Pluto was in some ways similar to the discovery of Uranus and Neptune and in some ways different. Analysis of perturbations in the motion of Uranus and Neptune by W. H. Pickering and Percival Lowell early in this century led them to predict the existence of a trans-Neptunian planet. A photographic search in 1919 by Milton Humason of the Mount Wilson Observatory would have found the planet had it not been for a double piece of bad luck. Many years later it was found that two of Humason's plates actually showed an image of Pluto, but it was hidden by a defect in one plate and obscured by a bright star in the other. Workers at the Lowell Observatory in Flagstaff, Ariz., which was founded specifically for planetary studies, mounted several surveys, none of them successful. Finally in 1929 a special telescope was built at the Lowell Observatory to find the trans-Neptunian planet, and in February, 1930, Clyde W. Tombaugh discovered the planet about five degrees from its predicted position. It now appears, however, that the observed mass of Pluto is far too small to produce the perturbations of Uranus and Neptune that originally led to the prediction of the planet's existence. In other words, the calculations had led to the discovery but they were wrong.

If it was difficult to find the outermost planets, the study of their physical nature is even more so. Except for the observations made from the two Pioneer spacecraft that have floated past Jupiter, physical studies of the outer planets have so far had to be made by astronomical methods. The exploration of space has nonetheless had strong indirect effects, and in the past 10 years there has been a renaissance of planetary science. The mere act of planning future space missions has stimulated an entire series of advances, both observational and theoretical.

Since the outer planets are so far from the sun they receive only a small amount of light and heat. The paucity of light re-

**PHOTOGRAPH OF SATURN** shown on the opposite page is composite of 16 images made by Stephen M. Larson with the 155-centimeter reflecting telescope at the Catalina Observatory in Arizona. The rings, which are directly over the equator, and the dark cap over the south pole are tilted 26.9 degrees to the line of sight. As viewed from the earth the tilt of the planet's axis changes, alternately presenting the southern hemisphere and the northern one every 30 years. The large dark gap in the rings is called Cassini's division. The visible surface of Saturn is yellowish, with darker belts that parallel the equator. The most prominent belt is near the equator. The shadow cast by the planet partly obscures the rings.

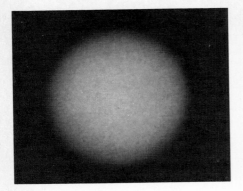

URANUS, photographed with a balloon-borne telescope at an altitude of 80,000 feet in the earth's atmosphere, shows no visible surface markings. If there are faint belts or markings on the planet, they have a maximum contrast of 5 percent. This photograph is a composite of 17 images taken during a flight of *Stratoscope II* in 1970. The images were combined with the aid of a computer at Princeton University by Robert E. Danielson, Martin Tomasko and Blair Savage.

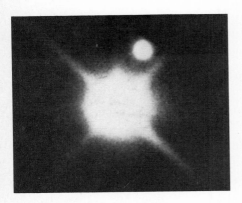

NEPTUNE appears in this photograph made with the 120-inch reflecting telescope at the Lick Observatory in California. Triton, the larger of Neptune's two moons, can be seen.

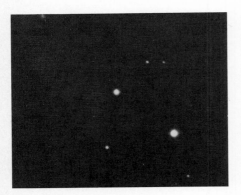

PLUTO, the outermost planet, appears in the center of this photograph made with the 200-inch telescope on Palomar Mountain. Pluto is so faint that it cannot be studied by customary visual and spectroscopic techniques. Photometric measurements, however, reveal that the brightness of Pluto varies about 20 percent every 6.39 days. Change in brightness is probably due to the planet's rotation and variations in surface features.

flected from them contributes to the difficulty of studying their features by telescope. Jupiter is the most easily observed of the outer planets, and it is clear that it has a deep atmosphere filled with clouds. Telescopic inspection of Saturn reveals that it too has a deep atmosphere with clouds. Some observers have reported that Uranus has faint bands, indicating that it has an extended atmosphere. Neptune is exceedingly difficult to observe even with very large telescopes, but there is some evidence that it has a deep atmosphere. Photometric observations indicate that the planet shows no periodic variation in brightness. Since it is almost impossible for a rotating body with a visible solid surface not to vary in brightness, the absence of such variation indicates that Neptune has an atmosphere. Pluto, on the other hand, does show a variation in brightness. It is presumably caused by the planet's rotation and by variation in its surface features. The brightness of Pluto varies about 20 percent with a period of 6.39 days.

Timing the occultation, or eclipse, of a star by a planet can provide accurate information about the size of the planet. The disappearance of the star behind the moving planet and its reappearance are usually measured with a highly sensitive photometer. A partial occultation by Pluto was observed in 1965, and the results indicate that the diameter of the planet cannot exceed 5,800 kilometers. (The diameter of the earth is 12,756 kilometers.) In 1968 excellent measurements of the occultation of a star by Neptune were made by astronomers in Japan, Australia and New Zealand, who obtained a new value for the diameter of the planet: 49,500 kilometers. Regrettably it has not been possible to observe any stellar occultations by Saturn or Uranus.

Observation of the most distant planets with even the largest telescopes is limited not only by the paucity of light reflected from the planets but also by the turbulence of the earth's atmosphere. One development that has been stimulated by the space program is the use of balloons to lift telescopes into the stratosphere in order to overcome the blurring caused by the lower atmosphere. In 1970 a balloon-borne telescope took a series of photographs of Uranus that had a resolution of .15 second of arc, about 10 times better than the best resolution that can be obtained with a telescope on the ground [see top illustration at left]. From these photographs Robert E. Danielson, Martin Tomasko and Blair

Savage of Princeton University determined that the diameter of Uranus is 51,800 kilometers.

Since the volume of a planet is proportional to the cube of its radius, accuracy in the measurement of the planet's diameter is of critical importance. A small change in the measured diameter or radius leads to a large change in the volume. The mean density of a planet, an important clue to its gross composition and internal structure, is obtained by dividing the planet's volume into its mass. The mass can be calculated from the orbital period of the planet's satellites or from perturbations in the orbits of nearby planets. Saturn, Uranus and Neptune each have moons that make it possible to accurately determine the planetary mass and the mean density. Pluto, however, has no satellites, and its mass seems to be too small to have any measurable effect on the orbits of its far more massive neighbors. That makes determining the mass of Pluto extremely difficult. The best estimate is that it has about a tenth the mass of the earth, but the probable error is actually greater than the estimate. Since the radius of Pluto is also not known with any great certainty, its mean density is not known, and the values usually quoted must be regarded as postulates [see top illustration on page 107].

It has been hard to determine the rotational period of Uranus and Neptune because the planets make such a small disk in a telescope and because visible surface features cannot be descried. Measurements of the Doppler shift of the spectral lines from the approaching and receding edges of each planet yield a rotation period of 11 hours for Uranus and 16 hours for Neptune, but the values are uncertain. Unlike any other planet, Uranus has a rotational axis that lies almost in the plane of its orbit. Careful studies of the surface features of Saturn have yielded a rotational period of 10.2 hours.

The atmospheric composition of the outer planets has been studied primarily by absorption spectroscopy and emission spectroscopy. The absorption spectrum of sunlight reflected from a planet provides clues to the presence and abundance of various gases, which absorb different wavelengths of light. The emission spectrum, on the other hand, provides evidence on the planet's own thermal emissions.

The atmosphere of Uranus is remarkably clear, and sunlight penetrates deep into it before being reflected. The absorption spectrum of Uranus displays strong bands from methane, indicating a

| | DISTANCE FROM SUN (ASTRONOMICAL UNITS) | SOLAR HEAT | TEMPERATURE (DEGREES K.) | | DIAMETER (KILOMETERS) | MASS | DENSITY (WATER = 1) | ROTATION PERIOD (HOURS) |
|---|---|---|---|---|---|---|---|---|
| | | | EQUILIBRIUM | ACTUAL | | | | |
| EARTH | 1 | 1,000 | 246 | 290 | 12,756 | 1 | 5.52 | 24 |
| JUPITER | 5.2 | 37 | 105 | 135 | 142,800 | 318 | 1.314 | 9.9 |
| SATURN | 9.54 | 11 | 71 | 97 | 120,000 | 95 | .704 | 10.2 |
| TITAN | 9.54 | 11 | 82 | 130? | 5,800 | .023 | 1.34 | 16 DAYS |
| URANUS | 19.2 | 2.7 | 57 | — | 51,800 | 14.6 | 1.21 | 11 |
| NEPTUNE | 30.1 | 1.1 | 45 | — | 49,500 | 17.2 | 1.67 | 16 |
| PLUTO | 39.4 | .64 | 42 | — | 5,800? | .1? | 2? | 6.39 DAYS |

**GENERAL PROPERTIES** of the outer planets are given, with those of the earth for reference. Titan, a satellite of Saturn, is included. A notable feature of the outer planets is the small amount of radiation they receive from the sun. Equilibrium temperatures are calculated from the amount of solar heat that is absorbed by the planet. Jupiter and Saturn radiate considerably more heat than they absorb, indicating they have an internal heat source. The actual temperature shown for the earth refers to its surface, which is warmed by the "greenhouse effect" of the atmosphere; the actual radiated energy very closely matches energy received from the sun.

high abundance of that gas, which may or may not be present in the form of clouds. Neptune is similar in having a clear atmosphere and in showing strong methane absorption. The spectrum of Saturn shows methane absorption, but the absorption is weaker than that for Uranus and Neptune [*see illustration on page 108*].

Ammonia bands, which are clearly present in absorption spectra of Jupiter, may or may not be present in spectra of Saturn and are not found in spectra of Uranus or Neptune. The probable explanation is that the lower temperature of Uranus and Neptune causes ammonia clouds to form at greater depths, where the clouds cannot be observed. Although Saturn does not show much ammonia absorption, the clouds visible in photographs of the planet are most likely to be ammonia.

Hydrogen is the most abundant gas on these planets, but it absorbs light very weakly and is difficult to detect in spectra. Hydrogen absorption is shown by Saturn and Uranus, and the gas is assumed to be present on Neptune. Helium does not absorb light at all, but it is presumably present in the same proportion in the outer planets as it is in the sun: about one atom of helium for every 10 molecules of hydrogen.

Although Titan is a satellite of Saturn, it is appropriate to discuss it here as a planet. It is larger than Mercury and almost as large as Mars. It has an atmosphere that is deeper than that of Mars. Strong methane absorption was found in Titan's spectrum by Gerard P. Kuiper of the University of Chicago in 1944. Recently Laurence M. Trafton of the University of Texas reevaluated Kuiper's results and found that the surface pressure of the satellite's atmosphere should be at least four times the pressure on Mars. If an invisible gas is

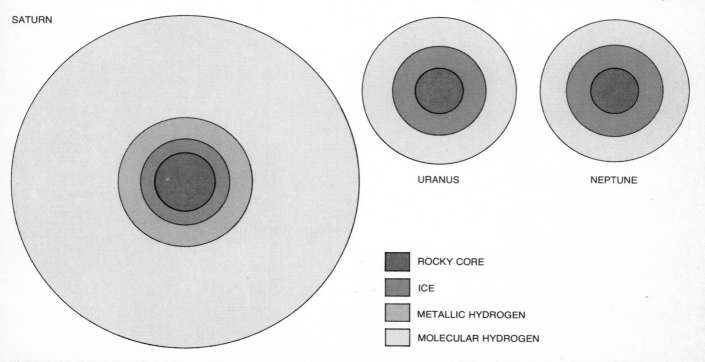

SATURN

URANUS

NEPTUNE

■ ROCKY CORE

■ ICE

■ METALLIC HYDROGEN

□ MOLECULAR HYDROGEN

**INTERNAL STRUCTURE** of Saturn, Uranus and Neptune is shown in cross section. These models are derived from the hypothesis that the planets were initially formed by the accretion of rocky material and ice, followed by the accumulation of gas. Saturn's rocky core, 20,000 kilometers in diameter, is surrounded by a 5,000-kilometer shell of ice and an 8,000-kilometer layer of metallic hydrogen. Uranus and Neptune each have a core 16,000 kilometers in diameter, surrounded by an 8,000-kilometer layer of ice.

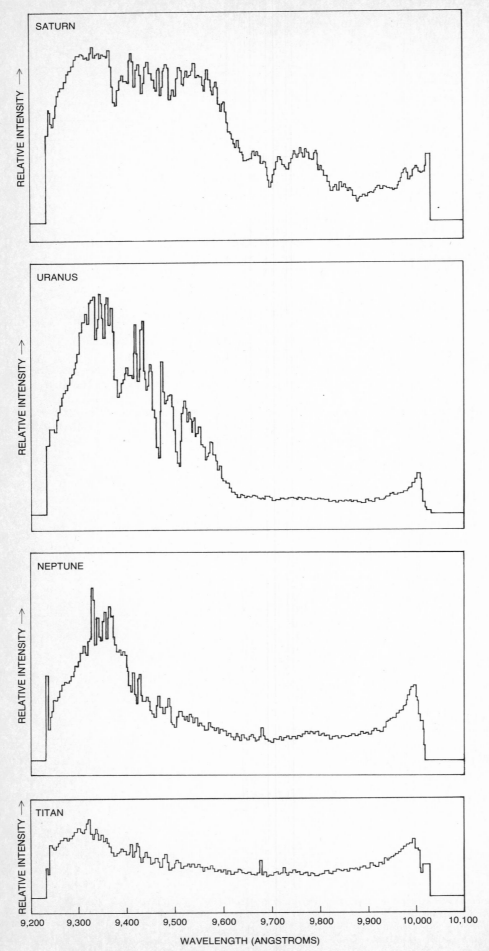

present along with the methane, a surface pressure as great as that on the earth is plausible. Following a suggestion by John S. Lewis of the Massachusetts Institute of Technology, I have proposed that the gas could be nitrogen formed from ammonia ($NH_3$) by the effect of radiation from the sun. Trafton also found evidence for hydrogen absorption on Titan, and hydrogen on the satellite may be as abundant as methane.

The upper levels of the earth's atmosphere are warmed by the absorption of solar ultraviolet radiation by the ozone present at those levels. A similar warming process seems to occur on Jupiter, Saturn, Uranus, Neptune and Titan, but the absorbing substances are different. One of the substances most likely to be an absorber is methane. L. W. Wallace and his colleagues at the Kitt Peak National Observatory have shown that absorption of solar energy by methane should raise the temperature of the upper atmosphere by 70 to 80 degrees Kelvin.

A second warming factor on Jupiter, Saturn and Titan is a "smog" of small dark particles suspended in the atmosphere. The presence of the smog is inferred from the fact that these bodies do not reflect ultraviolet radiation as well as they should considering the gases present. The source of the dark particles is not known, although they could possibly be formed by the linkage of methane molecules into polymers by solar radiation. The particles would absorb sunlight and transfer heat to the surrounding gas.

Gas molecules in a warm planetary stratosphere emit infrared energy that can be detected. The emission spectrum of Saturn reveals the presence not only of methane but also of a complex of ethane, ethylene and acetylene. The complex is probably the product of the sun-induced polymerization of methane.

There also is some evidence for a warm stratosphere on Neptune. The analysis of data obtained from the stellar occultation by the planet in 1968 suggests that the temperature of its upper atmosphere may be as high as 140 degrees K. The infrared emission spectrum of Titan is similar to that of Saturn, in-

INFRARED ABSORPTION SPECTRA of Saturn, Uranus, Neptune and Titan made recently with an experimental spectrograph on the 400-centimeter telescope at the Kitt Peak National Observatory are shown. Methane absorption, which is evident from 9,600 to 10,000 angstroms, is stronger on Uranus, Neptune and Titan than on Saturn.

dicating that Titan also has a warm stratosphere [*see illustration below*].

The rings of Saturn are among the most beautiful objects that can be seen in a telescope. More than a century ago James Clerk Maxwell concluded that the rings must consist of individual small objects in orbit around the planet. Further understanding of the nature of the objects has come only in the past few years. A reflection spectrum of the rings at near-infrared wavelengths was obtained in 1970. At first it was thought that the spectrum matched that of frozen ammonia, but it was quickly pointed out that it matched the spectrum of ice at very low temperatures much better. Radar reflections from the rings were obtained in 1972, and although the interpretation is still not entirely agreed on, the most convincing one is that the particles are ice and have diameters between four and 30 centimeters (between an inch and a half and a foot). Photometric observations of the rings under various conditions of illumination suggest that the surface texture of the objects is more like snow than like solid ice.

It has recently been realized that there should be rings of gas associated with some of Jupiter's and Saturn's satellites. Such bodies are not large enough to hold hydrogen permanently, as Jupiter and Saturn can. Therefore if hydrogen is present on Titan, much of it should escape from the satellite's atmosphere and go into orbit around Saturn. If the lifetime of the gas in orbit were long enough, it would form a ring centered on Titan's orbit. The ring would in effect be an extension of Titan's upper atmosphere. Estimates of the lifetime of the gas in orbit suggest that the density of the ring would be very small: 1,000 molecules or fewer per cubic centimeter. The spacecraft *Pioneer 10* sent back evidence that there are hydrogen atoms in the vicinity of one of Jupiter's 13 satellites, Io. The thinness of the hydrogen supports the idea that the ring in Titan's orbit would not be very dense.

As new facts have come to light during this century, concepts of the nature of the giant outer planets have gone through some remarkable swings. At one time these planets were regarded as being miniature suns, but measurements of the amount of heat they radiate dispelled that notion. In 1937 Rupert Wildt of Yale University proposed that Jupiter and Saturn, because of their large size and low density, must consist chiefly of hydrogen, with a large core of metal and rock enclosed in a mantle of ice. The view held today is that about 15 percent

of the mass of Jupiter and Saturn is a rocky and metallic core and the rest is mainly hydrogen and helium. It is likely that the interior is hot. Uranus and Neptune are denser and must therefore contain a larger proportion of rock and metal, perhaps as much as 90 percent.

In 1969 the infrared thermal emission of Jupiter and Saturn was measured from an aircraft flying at a high altitude in the earth's atmosphere. The findings showed that both planets radiate roughly twice as much energy as they receive from the sun. That discovery led to another revolution in our notions about the giant outer planets, and has brought about a moderate return to the idea that these planets have an internal source of heat. The total output of power from Jupiter, however, is only $3 \times 10^{-7}$ that of the sun, and Saturn's output is five times weaker still. The amount of thermal energy emitted from Uranus, Neptune and Pluto is still not known.

According to one hypothesis, the source of the energy emitted by Jupiter and Saturn is a slight, continuing contraction of the planets, which transforms gravitational energy into heat. Since much of the heat originates deep in the interior, it must somehow be transported out. The only process that has the necessary capacity is thermal convection currents, which must be present at great depths. If convection does operate, then the materials of the interior must be mixed rather than stratified. Models of Jupiter and Saturn in which the interior is fully mixed have been proposed by William B. Hubbard of the University of Arizona.

Another hypothesis is that Jupiter and

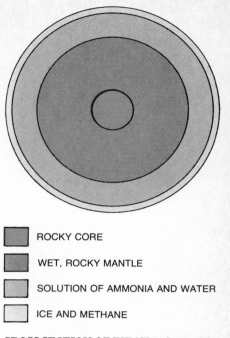

▇ ROCKY CORE

▇ WET, ROCKY MANTLE

▤ SOLUTION OF AMMONIA AND WATER

▢ ICE AND METHANE

**CROSS SECTION OF TITAN is shown. The core is thought to be surrounded by a rocky mantle with much water bound in the rock. The mantle may be covered with a "magma" of water containing dissolved ammonia. The crust may be a mixture of ice and methane.**

Saturn are simply radiating the heat they acquired during the gravitational-contraction phase of their formation, and that the planets are slowly contracting as they cool. This model also predicts large-scale thermal convection in the interior of the planets.

Concepts of the composition of the giant outer planets are closely related to modern hypotheses on the origin and

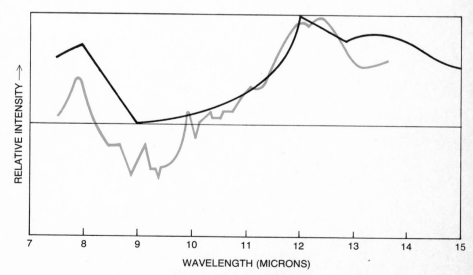

**INFRARED EMISSION SPECTRUM** of Saturn (*colored curve*) and that of its satellite, Titan (*black curve*), are compared. For Saturn, and probably for Titan, the radiation shown here is primarily thermal emission from the atmosphere. The peak at about eight microns is emission from methane and the peak at about 12 microns is possibly emission from ethane.

evolution of the solar system (see "The Origin and Evolution of the Solar System," by A. G. W. Cameron, page 15). The nebula that gave rise to the sun and the planets was formed by the collapse of an interstellar cloud. Much of the material in the solar nebula flowed inward to form the sun, but some of it remained in orbit. The giant outer planets began to take shape by the accretion of rocky material and ice. As the rocky cores materialized they began to draw to them large amounts of hydrogen and helium from the nebula. The process went furthest in the formation of Jupiter and Saturn, which swept up virtually all the gas in their vicinity. The process stopped relatively early for the more distant Uranus and Neptune, giving them a greater proportion of rocky material with respect to hydrogen and helium. Since the planets formed by accretion, they would have consisted of a series of shells [see bottom illustration on page 107]. Presumably the convection required to transfer heat from the interior proceeds separately within the shells because of their greatly differing densities. The material within each shell would be well mixed but there would be little mixing between shells.

Do any of the giant outer planets have a solid surface? The answer depends on what is meant by a solid surface. The model of Jupiter that arises from the accretion hypothesis has a rocky core, but the temperature of the core is nearly 20,000 degrees K. and the pressure is 60 million atmospheres. The material could be regarded as either a liquid or a highly compressed vapor. In the model of Uranus the core has a temperature of 4,000 degrees and the pressure is more than two million atmospheres. Whatever else we may say about such a surface, we can be sure that it is inaccessible. (Very little can be said about the structure of Pluto until we learn more about its mass and size.)

Titan's mean density is 2.1 grams per cubic centimeter, and it too is believed to have a metallic and rocky core. Around the core there may be a "magma" of water with ammonia dissolved in it. The crust may be ordinary ice with considerable amounts of methane trapped in it. If the atmosphere is deep, the ice crust could melt to leave a layer of liquid methane floating on the water-ammonia solution. Titan is a body unlike any other we know of. It is neither like one of the inner planets nor like one of the giant outer planets but is a kind of hybrid. Its close exploration cannot fail to be highly rewarding.

# THE SMALLER BODIES
# OF THE SOLAR SYSTEM

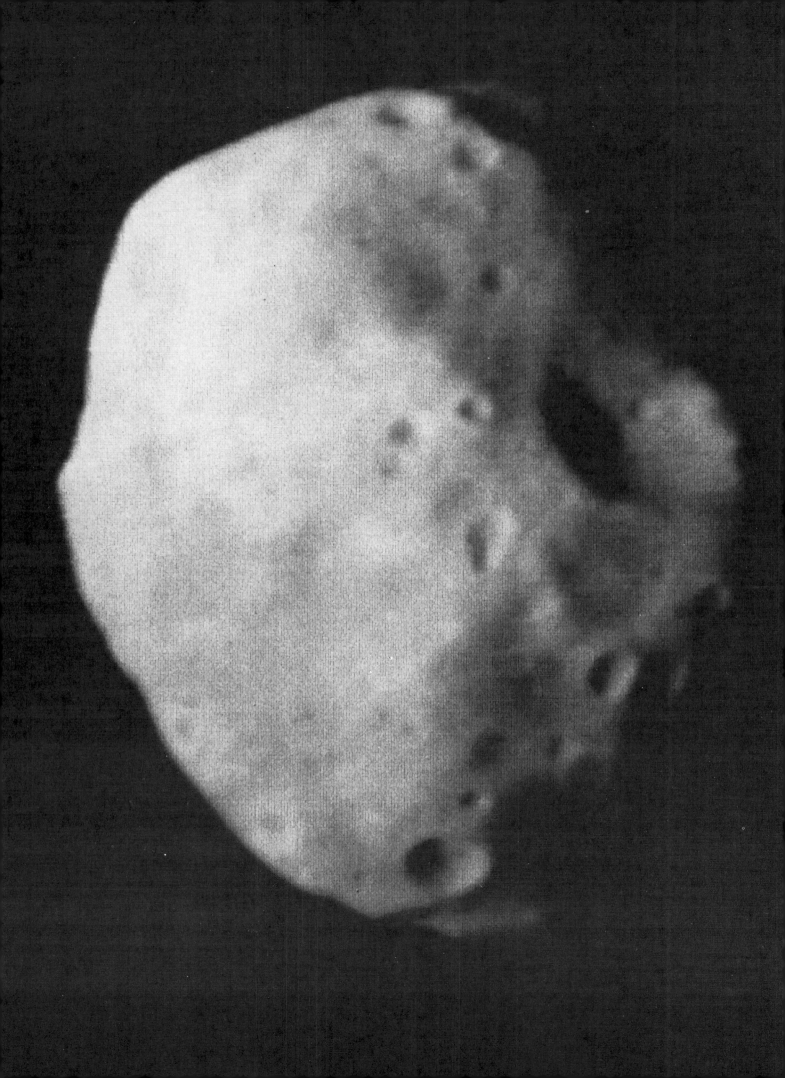

# The Smaller Bodies of the Solar System

WILLIAM K. HARTMANN

*They range in size from meteoroids no larger than a grain of sand to moons bigger than the planet Mercury. Many of them appear to be fragments resulting from collisions between growing planetesimals*

Most people are surprised to learn how many planetlike objects there are in the solar system. It harbors four bodies whose diameters are probably greater than the diameter of Mercury and yet are not listed among the nine planets. Eleven other nonplanets are probably larger than any asteroid. These 15 bodies are the satellites of various planets. All told there are 33 satellites, shared by six of the nine planets; they range in diameter from a few kilometers to about 6,000 kilometers. There are in addition some 2,000 sizable asteroids with fairly well-known orbits; the largest of them is 955 kilometers in diameter. Many more thousands of small asteroids, a kilometer or less in diameter, follow paths around the sun that are less well known. Furthermore, billions of comets, their diameters measurable in tens of kilometers, are believed to travel in orbit beyond the outer planets. Here we shall consider how all the lesser bodies are related: the satellites, the various groups of asteroids, including meteoritic rubble ranging down to the size of a grain of sand, the icy comets, such odd debris as the reflective chunks that make up the rings of Saturn and even the dust grains that condensed before the planets began to grow.

Until a few years ago the smaller bodies of the solar system received little attention. That attitude is changing rapidly. The amount of research that has been published on the subject has increased by a factor of 10 since 1960.

Early in the 1970's entire sessions of various astronomical meetings were devoted to papers about the physical nature, the origin and the surface properties of the lesser bodies. In 1971 a five-day conference at the University of Arizona dealt with the asteroids, and in 1974 satellites were the topic of a three-day symposium at Cornell University.

The reasons for the trend are not hard to find. Many of the lesser bodies were discovered more than a century ago, but the telescopes and spectrographs of the time were not sensitive enough to enable the observers to learn much more about them than their orbits. During the first half of this century the burning astronomical issues were nonplanetary: the origin of the energy of the stars, the structure of the galaxy and the nature of the universe as a whole. Why be concerned about a few cold little worlds in our backyard?

During World War II the technology of infrared spectroscopy advanced rapidly, making possible the study of that part of the electromagnetic spectrum where the absorption bands of many cool gases appear. In the same region one can observe subtle spectral features that arise from the mineralogical properties of rocks on extraterrestrial bodies. It was this development that enabled Gerard P. Kuiper of the University of Chicago to discover in 1944 the first nonplanetary atmosphere in the solar system: the mantle of methane gas that envelops Titan, the largest moon of Saturn. For many

years thereafter, however, the application of even the best instruments led to findings that were hardly earthshaking: some of the lesser bodies were pinker than others, and so on. Exciting progress began when interest and financial support, particularly on the part of the National Aeronautics and Space Administration in the 1960's, led to the development of a number of improved instruments and new ways of utilizing them.

Preparation for the Apollo program led to intensive studies of the moon and moon-related problems. Geochemical laboratories that were eventually to analyze lunar rock samples practiced on meteorites, the only rocks from interplanetary space that were then available. Studies both before and after the Apollo landings led to a burgeoning of interest in all moons. Astronomers undertook sophisticated observations of the spectral properties of planetary satellites and engaged in new investigations into the origin of satellites, asteroids and comets.

In 1971 interplanetary probes relayed to the earth closeup observations of the two tiny satellites of Mars. More distant views of Jupiter's giant moons were obtained in 1973. Two probes have passed through the asteroid belt, and the manned spacecraft have brought samples from the moon for laboratory analysis. As Paul D. Lowman, Jr., a NASA geologist, has noted, in all of geology's 200-year history as a science the most intensively studied specimens are not terrestrial rocks but lunar ones.

One result of all this research is a growing recognition of two largely unanticipated findings. First, the lesser bodies of the solar system provide important informational links between the present and the past. Second, the data establish connections between two hitherto largely separate disciplines: planetary science and astrophysics. Some of

ASTEROIDLIKE BODY seen in the photograph on the opposite page is the larger moon of Mars, Phobos. Measuring 20 by 23 by 28 kilometers, it has been cratered by numerous collisions with smaller bodies. Neither Phobos nor its companion, Deimos, is in the kind of orbit around Mars that would be expected of a captured asteroid. The two satellites may be the remains of a shattered larger satellite that was destroyed by impacts of the kind that have scarred Mars. The photograph was made from *Mariner 9* while it was in orbit around Mars.

those linkages will be discussed here. Starting with an inventory of facts about the lesser bodies, considered in order outward from the sun, I shall attempt some systematization of the data by examining the distribution of mass among all the bodies in the solar system. Finally, I shall suggest how smaller bodies might have originated and evolved in the kind of cloud that gave rise to the rest of the solar system.

The first bodies outward from the sun are Mercury and Venus. Neither planet has any satellites, but there is a good reason for mentioning them here. The craters on their surface show that many bodies as large as 100 kilometers in diameter have collided with them. At present only a few small comets and asteroids with highly eccentric orbits visit that innermost part of the solar system. Are those occasional visitors remnants of a larger population of planetesimals that inhabited the inner solar system long ago? Does the cratering indicate that collisions like these were an important factor in the evolution of the solar system?

Recent studies of our own satellite, the moon, have begun to provide affirmative answers to these questions. Lunar science is reviewed in detail elsewhere in this issue [see "The Moon," by John A. Wood, page 69], but let us list some key facts here. The moon is unusually large compared with its planet. It is made of materials that somewhat resemble the materials of the earth's surface layers but differ from the materials of its interior. This composition has never been adequately explained. D. R. Davis and I have recently proposed that the impact of a huge body may have knocked loose from the outer layer of the earth much of the present lunar mass. Initial heating of the moon led to partial melting and the formation of a primeval crust. In the first few hundred million years of its existence that crust was intensely bombarded by planetesimals, some of them more than 100 kilometers in diameter; as a result huge craters crowded the lunar surface. The intense bombardment declined about four billion years ago. Thereafter some parts of the moon's cratered primeval crust were melted by further heating and other parts were inundated by lava flows. Since then the moon has been further bombarded, but the bombardment, although relatively constant, has been low in intensity.

Small bodies still bombard the inner planets. For example, one short-lived group continues to cross the orbit of the earth, leaving open the possibility that one or another of them may one day crater the earth or the moon. These are the Apollo asteroids, a group that was named after one of its members. James Williams, Brian G. Marsden, Eugene M. Shoemaker and Fred L. Whipple are among those who are studying and tabulating the asteroids of this group. A recent listing includes 19 Apollos that range in diameter from about 200 meters to as much as six kilometers. Observers believe many more remain to be discovered. The risk to the inner planets is not confined to the Apollos. There is a great reservoir of other asteroids, and a greater reservoir of comets, that could at any time be deflected into orbits intersecting the orbit of the earth.

The Apollo asteroids have been studied in many different ways. Two of them, Eros and Icarus, have been probed by radar. In January of this year the former astronaut Brian O'Leary directed a novel Apollo-asteroid project. On the night of January 23, according to calculations that were completed only two hours before the event, Eros was to eclipse a third-magnitude star in the constellation Gemini, and the asteroid's shadow would sweep across Connecticut in the general vicinity of New Britain. Armed with predictions of the shadow path, teams of observers raced to various sites along the path and timed the three-second occultation of the star. Eight positive observations, and several negative ones from points outside the shadow path, made it possible to recalculate the size and shape of Eros: it is a slab that measures about seven by 19 by 30 kilometers.

The Apollo asteroids have proved to be related to a much more familiar group of lesser bodies that intersect the earth's orbit: the meteorites. The similarity of the orbits of both groups has been noted for some time. Comparisons of spectral characteristics have recently provided a second similarity. Clark R. Chapman, Thomas B. McCord and their associates recently undertook spectrophotometric studies of many different asteroids and compared their results with the laboratory spectra of five separate classes of meteorites. They found a number of good matches; the closest match was between the Apollo asteroid Toro and the class of meteorites known as L-type chondrites. Perhaps some of the L-type chondrites in museums around the world are pieces knocked loose from Toro. And perhaps as such studies continue it will be possible to identify other individual asteroids as the source of specific kinds of meteorites.

As for the meteorites themselves (or meteoroids, as they are called while they are still in space), they exist in abundance but are detected only when they graze the earth's atmosphere, burn up in it or survive to reach the ground. Fist-size meteoroids are so common that several strike the earth every year. Today meteoroid data are sometimes collected by Air Force monitoring devices that are designed to detect long-range missiles in flight, but most such information is kept secret. A widely observed incident in 1972, however, provides an example. On April 10 of that year both Air Force satellites and people on the ground observed the passage of a meteoroid at an altitude of 60 kilometers over the state of Montana. The meteoroid, which according to one estimate weighed perhaps 1,000 tons, skipped off the earth's atmosphere back out into space. If such a body had been a few kilometers lower, it might have crashed in Alberta, producing a tremendous explosion and a crater perhaps 100 meters in diameter. It is not known how often there are such major near-misses or even direct hits. Some analysts have suggested that the rate is about one collision with a body considerably larger than the Montana meteoroid per century. (It is a comforting thought that three in four large meteorites on a collision course with the earth would fall in the oceans rather than on land.) One such collision, probably with a comet, took place in Siberia in 1908; it was observed by a few stunned witnesses. If the same body had hit New York, the scorched area would have reached beyond Jersey City, people could have been knocked unconscious all the way from Philadelphia to Hartford and the "deafening bang" reported by Siberian witnesses might have been heard from Washington to Boston and possibly as far away as Pittsburgh. Meteoroids and Apollo asteroids are not entirely academic concerns.

Continuing outward from the sun, Mars is found to be attended by two peculiar moons that are no larger than modest-sized asteroids. The remarkable photographs made by *Mariner 9* provide further evidence of interplanetary collisions. Both chunky moons are liberally cratered. Phobos is 20 by 23 by 28 kilometers and Deimos is only 10 by 12 by 16. Their circular orbits around Mars are not of the kind one would expect if they were true asteroids, gravitationally captured by the planet from the main asteroid belt nearby. Several of my colleagues and I have therefore proposed that they may be the shattered and perhaps partially reconstituted remains of an earlier single large satellite of Mars.

The main asteroid belt, a collection

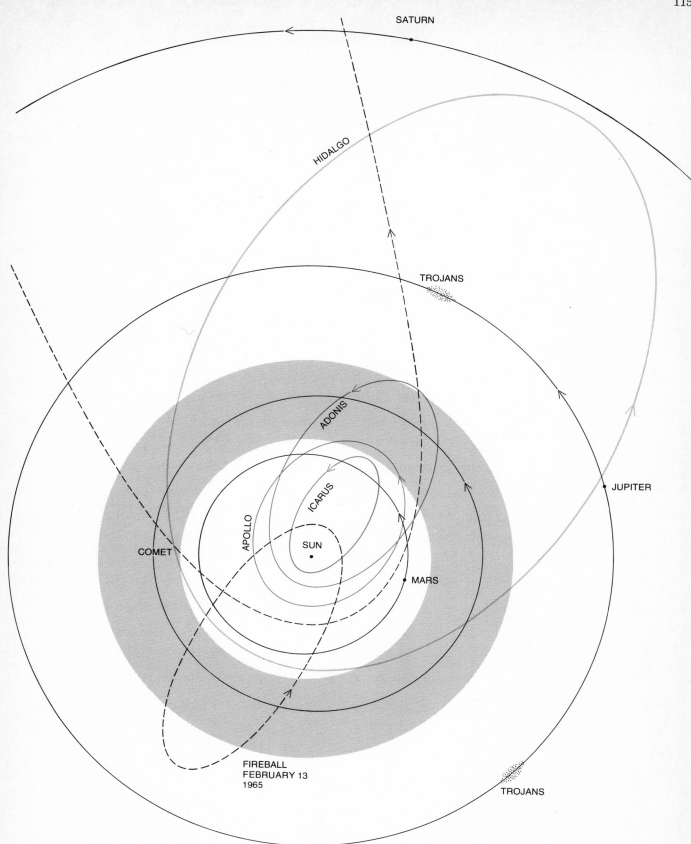

SATURN

HIDALGO

TROJANS

ADONIS

ICARUS

APOLLO

COMET

SUN

MARS

JUPITER

FIREBALL
FEBRUARY 13
1965

TROJANS

**ASTEROIDS** circle the sun in a wide variety of orbits. Some representative orbits are seen here in a view from above the plane of the solar system. Most of the asteroids are in a belt (*gray*) between Mars and Jupiter; variations in their concentration are illustrated on the next page. The Apollos, 19 small asteroids that are named after one member of the family, follow unusual orbits that carry some of them inside the orbit of Mercury; three such orbits are shown (*color*). A family of perhaps 1,000 asteroids, known as the Trojans, lies in the orbit of Jupiter (*black*). Twice as many asteroids lie at the Lagrangian point 60 degrees ahead of Jupiter as lie 60 degrees behind. The most eccentric of all asteroids is Hidalgo (*light color*); at its farthest from sun Hidalgo approaches orbit of Saturn. Also shown are orbits of other small bodies (*broken lines*): a comet and a meteoroid that grazed earth's atmosphere in 1965.

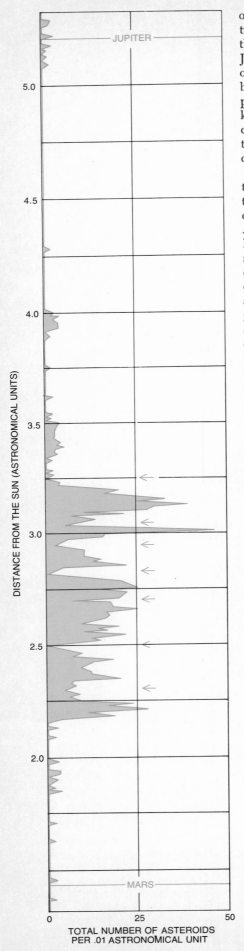

DISTANCE FROM THE SUN (ASTRONOMICAL UNITS)

TOTAL NUMBER OF ASTEROIDS
PER .01 ASTRONOMICAL UNIT

of thousands of rocky bodies, lies between Mars and Jupiter. The majority of the asteroids are closer to Mars than to Jupiter, which accords with Bode's law of planetary spacing. They are probably bodies that never accumulated into a planet. The largest asteroid is Ceres, 955 kilometers in diameter. It has been calculated that the combined mass of all the other main-belt asteroids would be close to the mass of Ceres.

Recent spectrographic studies show that the asteroids are divisible into distinct classes [see "The Nature of Asteroids," by Clark R. Chapman; SCIENTIFIC AMERICAN, January 1975]. Nearly 200 large asteroids have now been examined spectrophotometrically. Roughly 10 percent show spectra that resemble the laboratory spectra of the meteorites classed as stony irons. Asteroids of this class are generally 100 to 200 kilometers in diameter and are concentrated in the part of the asteroid belt nearest Mars. The spectra of some 80 percent, however, resemble the spectra of the most primitive meteorites known: the carbonaceous chondrites. It now seems firmly established that carbonaceous chondrites contain minerals representative of the ancient materials that formed when the nebula that gave birth to the sun began to cool. They are the least heated and least metamorphosed of all the meteorites and still contain abundant water in various chemical combinations.

How are these asteroid-meteorite resemblances to be interpreted? The asteroids that resemble carbonaceous chondrites may be those that were never fragmented by collisions or were never big enough to get very hot inside. Hence they have retained their primitive mineral surface. The asteroids that resemble the stony irons may be the remnants of the metallic cores that formed inside larger bodies when heating and melting led to the segregation of light and heavy materials. What we now see is widely believed to be the end product of catastrophic collisions that chipped away the larger bodies' outer mantle of light silicates. Ceres, which is nearly twice the diameter of the second-largest asteroid, Pallas, may be a more advanced planetesimal, one that had begun to approach planetary dimensions just as the main formative processes in the solar system came to an end.

Still other asteroids lie beyond the main belt, locked into stable positions along the orbit of Jupiter 60 degrees ahead of the giant planet and 60 degrees behind it. The two clusters are called the Trojan asteroids, and about 1,000 Trojans are now known. They occupy the two points in the sun-Jupiter system where Joseph Louis Lagrange showed that small bodies might accumulate. (It was not until 1905, almost a century after Lagrange's death, that Achilles, the first of the Trojan asteroids, was discovered.) Spectral studies show that as a group the Trojans are the darkest of all the asteroids. They may be composed of debris left over after the formation of Jupiter, or they may be accretions of interplanetary material gravitationally attracted toward the giant planet. Recent studies by Tom Gehrels of the University of Arizona show that for some unknown reason there are at least twice as many Trojans at the Lagrangian point ahead of Jupiter as there are at the point behind it.

Jupiter itself has more satellites than any other planet in the solar system. The total is now 13, the 13th having been found only last year by Charles T. Kowal of the California Institute of Technology. Jupiter's satellites fall into three groups. The innermost group numbers five. The closest to the planet is a small moon, Amalthea. Then come the four large moons Galileo discovered; in outward order they are Io, Europa, Ganymede and Callisto. Amalthea is so small, perhaps only 150 kilometers in diameter, that little is known about it. The orbits of all five inner moons are circular and lie in the plane of Jupiter's equator.

Io, a good-sized body with a diameter of 3,640 kilometers, displays some of the most bizarre phenomena found in the solar system. At the turn of the century an astute observer, Edward Emerson Barnard of the Yerkes Observatory, reported the first of them: the poles of Io have reddish caps. Barnard's finding has now been documented in color photographs by R. B. Minton of the University of Arizona. Further peculiarities were reported in the 1960's by A. B. Binder and Dale P. Cruikshank, who were then at the University of Arizona. They found that Io, which frequently enters Jupiter's vast shadow, is at least a few percent brighter than usual for 15 minutes or so after it emerges. Binder

**ASTEROID BELT** (*left*), a region that starts about 2.2 astronomical units away from the sun, displays an unevenness in distribution of bodies. The seven major depressions in the graph between 2.2 and 3.3 A.U., known as the Kirkwood gaps, are spaces where few asteroids are present because of perturbations due to Jupiter's gravitational field.

and Cruikshank suggested that Io might have an atmosphere and that during the cold, dark eclipse clouds or frost might condense, to dissipate when the moon emerged into the sunlight.

Since then many more eclipses of Io have been observed. It has been found that the brightening takes place only about half of the time. Why that is so remains a mystery. Recent evidence suggests, however, that the phenomenon may be due not to clouds or frost but to the alteration of colored materials on the surface of Io, perhaps compounds of sulfur, by low temperatures during the eclipse.

Another strange thing about Io was noted in the 1960's by the radio astronomer E. K. Bigg, who observed a strong correlation between the position of Io in its orbit and bursts of radio noise from Jupiter. Apparently the radio noise is generated when Io disturbs the magnetic field of Jupiter in its swing around the planet.

In 1973, during the approach of the spacecraft *Pioneer 10* to Jupiter, Arvydas J. Kliore and his colleagues at the Jet Propulsion Laboratory of the California Institute of Technology detected a layer of ionized particles about 100 kilometers above the surface of Io. The observation means that the satellite has an atmosphere, albeit an extremely sparse one. Its surface pressure is estimated to be as low as one billionth the atmospheric pressure at the surface of the earth. In 1974 Robert Brown of Harvard University added to the list of oddities. He discovered that Io was surrounded by a yellow glow: the *D*-line emission characteristic of sodium. The glow extends outward from Io for several diameters. It is caused by the resonance scattering of sunlight by sodium atoms on the surface of the satellite.

Brown's discovery represents a pleasing convergence of theory and observation. Some time ago Fraser P. Fanale, a planetary chemist at the Jet Propulsion Laboratory, proposed that Io was covered by sodium-rich evaporite salts. If Io had been heated and its water had been driven off, Fanale reasoned, the salts formerly dissolved in the water would have been left behind on the surface. When all these observations are put together, the model that best fits Io's peculiar properties is a reddish brown world, covered with evaporite crystals (perhaps crystals of common salt), with darker reddish sulfur deposits at the poles and a sky often aglow with a yellow aurora. Overhead, on the side of Io facing Jupiter, the disk of the planet would measure 18 degrees from side to

side (about the same width as the Big Dipper).

The other three Galilean satellites of Jupiter have surfaces covered in varying degrees not so much by salt and sulfur as by gravelly soil and frost. Europa, 3,100 kilometers in diameter, is smaller than Io. Ganymede, 5,270 kilometers in diameter, is one of the largest satellites in the solar system and Callisto,

at 5,000 kilometers, is only a little smaller. Callisto is also darker than the others, apparently because more rocky material is exposed. The chemical differences between the four satellites remain the subject of current research.

Next outward from Jupiter is a group of four satellites ranging from 18 to 60 kilometers in diameter. They travel in orbits inclined at an angle of about 28

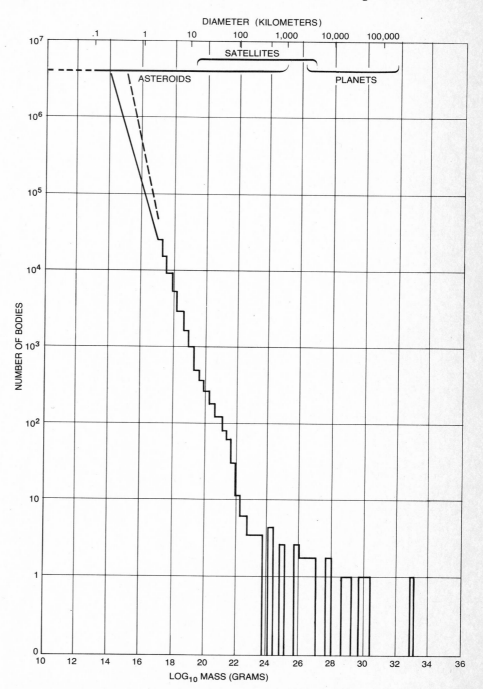

**DISTRIBUTION OF MASS** among the bodies of the solar system is plotted in this logarithmic graph. At the small-mass end of the graph (*left*) there is uncertainty about the number of bodies in the system; the broken line shows the effect of including comets. The parts of the curve at the center and the right are based on actual counts. It is evident that a disproportionate share of the total mass is possessed by the planets and their satellites. This suggests that the larger bodies grew by sweeping up most of the initial supply of small planetesimals. Those planetesimals that were not incorporated into larger bodies are very numerous. When compared with the planets, however, the leftover planetesimals are small.

degrees to the planet's equatorial plane and about 11.6 million kilometers from the planet. One of the four is Kowal's 1974 discovery, satellite No. 13.

The outermost group of Jupiter's satellites also numbers four. All are small: they range from about 16 to 22 kilometers in diameter. The four are about 22 million kilometers from Jupiter and are notable because, unlike the other nine satellites, they circle it in a retrograde direction. Their orbits are inclined at an angle of about 25 degrees to the planet's equatorial plane.

The two groups are puzzling. The clustering of orbits in each group, the high inclination of those orbits and the retrograde revolution of the outer group all indicate that these satellites are unlike the miniature solar system formed by Jupiter and the five inner satellites. Perhaps the two outer groups are plane-tesimals captured by special processes that favored each new captive's joining either one group or the other. Or perhaps, as the Russian worker Z. A. Aitekeeva suggests, the two groups are the clustered fragments of two larger satellites that were shattered by collisions with comets or with errant Trojan asteroids. Here it is worth remembering that close interactions and catastrophic collisions were probably rather common

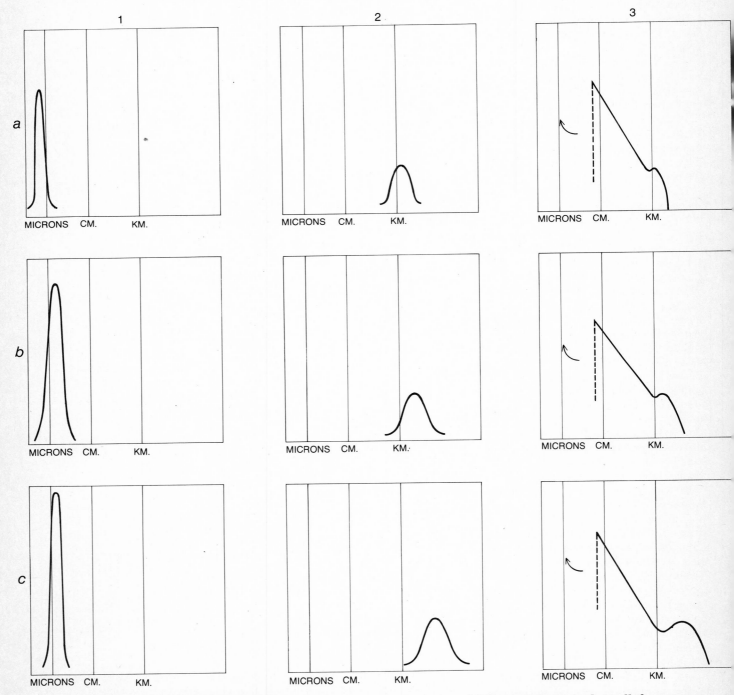

EVOLUTION OF SMALL BODIES from the nebular condensates in three stellar systems is traced in a sequence of schematic size-distribution diagrams. At first (1) dust grains that have formed in the three systems are all micron-size. The abundance of the grains differs, however; System A contains the least dust, System C the most and System B is intermediate. Gravitational accumulation (2) produces multikilometer planetesimals in all three systems; they tend to be larger in System B and System C than in System A. Once the planetesimals are formed competition begins (3) between accretion and destructive collision; all three systems begin to lose any collision fragments less than a centimeter in diameter. Meanwhile low-speed collisions that increase mass and gravita-

events during the early days of the solar system, when planetesimals were much more abundant.

Saturn, the second-largest planet, also has the second-largest collection of satellites: a total of 10. The 10th was discovered in 1966 by the French astronomer Audouin Dollfus; it is Janus, the innermost of Saturn's moons, some 200 kilometers in diameter. Outward from the planet we find five more smallish moons before reaching the orbit of great Titan. The five range in diameter from 400 to 1,600 kilometers. They are probably icy bodies; their estimated densities fall between one gram and 1.5 grams per cubic centimeter. (The density of water ice is one gram per cc.)

In the years since Kuiper discovered Titan's atmosphere it has produced a string of surprises. The satellite has long been notable for its size. Early this year, however, Joseph F. Veverka and his associates at Cornell recalculated the diameter of Titan; they put it at about 5,800 kilometers. That places it among the three largest satellites in the solar system; perhaps it is the largest.

Veverka's group and other observers have found that what one sees through the telescope is not Titan's unobscured surface but a blanket of reddish brown

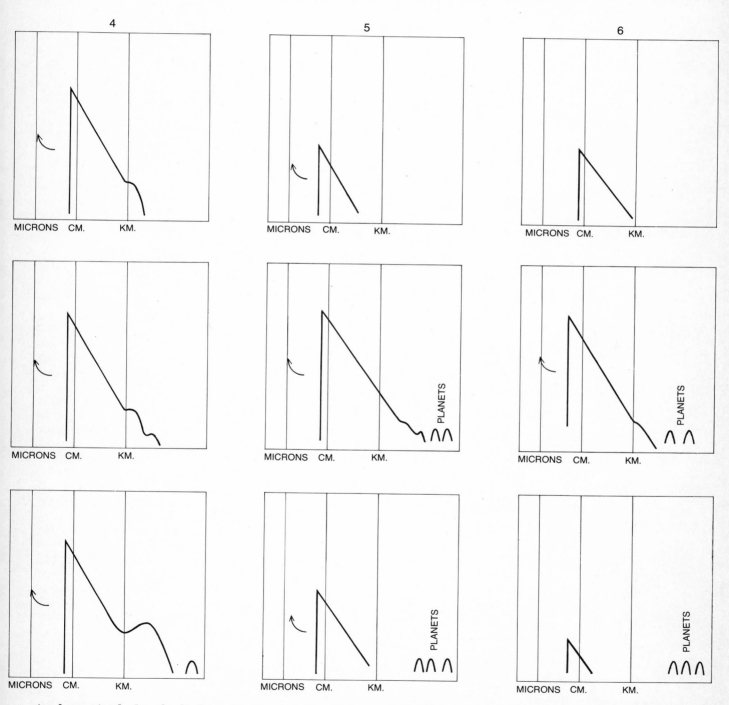

tional attraction lead to the development of larger bodies (4) in System *B* and System *C*. In System *A*, however, none of the planetesimals is large enough to continue growing, and collision overbalances accretion. In due course (5), although accretion has produced planet-size bodies in System *B* and System *C*, the planetesimals in System *A* are all smaller than kilometer-size. In System *C* all the kilometer-size bodies and almost all the smaller collisional debris have been swept up by the growing planets (6). Because planets did not grow in all parts of System *B*, many kilometer-size planetesimals were not swept up, and smaller collisional debris is also plentiful. The small-body populations in System *B* resemble the asteroids and the other, lesser debris found in the solar system.

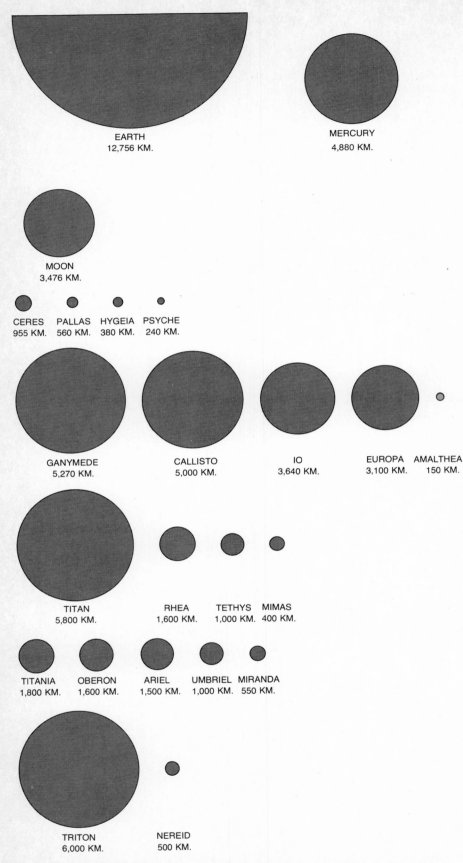

EARTH
12,756 KM.

MERCURY
4,880 KM.

MOON
3,476 KM.

CERES
955 KM.

PALLAS
560 KM.

HYGEIA
380 KM.

PSYCHE
240 KM.

GANYMEDE
5,270 KM.

CALLISTO
5,000 KM.

IO
3,640 KM.

EUROPA
3,100 KM.

AMALTHEA
150 KM.

TITAN
5,800 KM.

RHEA
1,600 KM.

TETHYS
1,000 KM.

MIMAS
400 KM.

TITANIA
1,800 KM.

OBERON
1,600 KM.

ARIEL
1,500 KM.

UMBRIEL
1,000 KM.

MIRANDA
550 KM.

TRITON
6,000 KM.

NEREID
500 KM.

**SELECTED SMALLER BODIES** are compared with the earth, which is shown only in part (*top left*), and Mercury (*top right*). Assuming that the diameters of Triton, one of the moons of Neptune, and Titan, one of the moons of Saturn, are correctly estimated, four moons are larger than Mercury. Ceres, the largest asteroid, is larger than several of the smaller satellites. For example, Amalthea (*color*), innermost of the 13 satellites of Jupiter, is estimated to be 150 kilometers in diameter. Not shown are 16 satellites: eight satellites of Jupiter, six of Saturn and two of Mars; if drawn to this scale, some would be merely dots.

clouds that conceals the ground, sometimes partially and sometimes totally. The atmosphere of Titan has also proved to be much denser than was once thought: it is primarily methane and hydrogen in about equal abundance.

Studies by James B. Pollack of the Ames Research Center of the National Aeronautics and Space Administration indicate that the atmospheric pressure at the surface of Titan is between .1 and one terrestrial atmosphere (the pressure of air at the earth's surface). The surface conditions on Titan, however, are unearthly. The surface itself may be ice. The temperature is about 125 degrees Kelvin, or 234 degrees below zero Fahrenheit. It may be, however, that Titan is not as unearthly as it might seem. Most geochemists believe the original atmosphere of the earth was rich in methane and hydrogen. If there is geothermal or volcanic activity on Titan, could not temporary pools of warm water exist? Under such circumstances the formation of complex organic molecules would be possible; conditions like these are precisely the ones that result in the synthesis of amino acids in the laboratory. Remote Titan thus holds some biochemical interest, although life there remains unlikely.

Two of the three satellites of Saturn that lie outside the orbit of Titan are unusual. Phoebe, the outermost, is a small moon that measures perhaps 250 kilometers in diameter. Its orbit, both retrograde and inclined, is reminiscent of the outer moons of Jupiter.

Phoebe's neighbor, Iapetus, has a most unusual appearance: one of its sides is six times brighter than the other. The leading hemisphere of Iapetus reflects only about 4 percent of the sunlight that reaches it; such low reflectivity means that it is darker than a blackboard. The trailing hemisphere, however, is highly reflective: it is about as bright as snow. A number of explanations have been offered. Did a major impact throw out fine bright dust (like the pulverized material blasted from large craters on the moon) that came to rest only on one part of the satellite? Did a swarm of meteorites pepper one part of Iapetus only, leaving a reflective ice surface undisturbed elsewhere? A particularly plausible hypothesis was proposed last year by Steven Soter of Cornell. He pointed out that the Poynting-Robertson effect (a net inward motion of small orbiting particles that are exposed to radiation) would make any fine dust that was eroded from Phoebe spiral inward toward Saturn. Like Phoebe itself, the dust particles would travel in a direction opposite to the motion of Iapetus, so that those intersecting the orbit of Iapetus would strike the leading hemisphere at high speeds. Soter suggests that this unique circumstance may be the source of Iapetus' unique visual asymmetry: perhaps the rain of particles causes abnormal erosion or dust accumulation or both.

Saturn is best known, of course, not for the curious properties of its satellites but for its spectacular rings. A backyard telescope makes the rings look like a solid disk around the planet, but as early as 1859 James Clerk Maxwell offered theoretical proof that the disk could not be solid, and in 1895 an American astronomer, James Keeler, confirmed Maxwell's prediction by means of spectroscopic observations. The rings are composed of innumerable thousands of satellite particles. A recent review of ring observations by Pollack suggests that the particles are typically a few centimeters in diameter.

The rings are divided by gaps that were probably created by dynamic interactions with the nearby satellites of Saturn. The rings are about 270,000 kilometers in total width but are only a few kilometers thick. A spacecraft passing through the plane of the ring system at a modest angle would encounter the ring particles for no longer than a few hundredths of a second.

Theorists are still struggling to explain the origin of Saturn's rings. Spectroscopic studies by Carl Pilcher and others show that the ring particles are composed of or covered with water ice. They may have condensed where they are during the formation of the solar system or they may be the fragments of pulverized satellites. Some explanatory models propose that the fine particles rarely collide today. Other models suggest that the rings are still evolving as a result of continuing collisions and that they do not necessarily date back to the origin of the solar system. From either viewpoint Saturn's rings are a natural laboratory for the study of swarms of small bodies in space.

In view of the oddities we have encountered so far, what is one to expect of the satellites of Uranus and Neptune? Because they are so distant very little is known about them. Uranus has five satellites; Neptune has two, one of which may be the largest satellite in the solar system. The five satellites of Uranus travel around the planet in the plane of its equator. The orientation is normal enough, except that Uranus itself is tipped at an angle of roughly 90 degrees to the plane of the planet's orbit

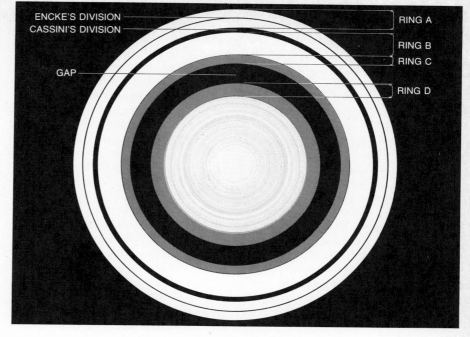

RINGS OF SATURN, shown schematically in a polar view, are made up of some of the more enigmatic small bodies in the solar system. Centimeter-size, they are either bits of ice or particles covered with ice. They are considered by some to be tiny planetesimals that condensed where they now are at the time the solar system was formed. Others hold them to be the debris formed by the collisional or tidal fragmentation of larger planetesimals.

around the sun. Evidently the force that tilted Uranus, whatever it was, either shifted the satellites along with the planet or acted on the planet before it had acquired satellites. A group of Russian astronomers and S. Fred Singer of the University of Virginia have independently proposed that what tilted Uranus was a collision with a planet-size planetesimal. The impact may have triggered the formation of the five satellites.

The inner satellite of Neptune, Triton, with an estimated diameter of 6,000 kilometers, is another satellite that moves in a retrograde orbit. This fact may be a clue to an unresolved puzzle. Pluto, with an estimated diameter of 5,800 kilometers, is usually given as the ninth planet in the solar system, even though its uniquely eccentric orbit sometimes brings it closer to the sun than Neptune ever gets. Some theorists, notably Kuiper and R. A. Lyttleton of the University of Cambridge, have suggested that Pluto was once a satellite of Neptune. Their explanatory models visualize an interaction between Triton and Pluto that was sufficiently forceful both to put Triton into its retrograde orbit and to eject Pluto from Neptune's sphere of influence. Cited as an objection to the theory is the fact that in spite of the eccentricity of its orbit Pluto does not approach Neptune closely today.

Although the orbit of Pluto lies some 40 astronomical units from the sun (40 times the distance from the sun to the earth), the census of the smaller bodies of the solar system is still far from complete. Traveling in the outer reaches of the system, sometimes thousands of astronomical units beyond Pluto, are billions of icy bodies with estimated diameters that range from 100 kilometers to a few tens of meters. These are the comets. Their orbital inclinations are virtually random. Occasionally one of them is perturbed by stellar or planetary forces and enters the inner solar system. Once a comet comes inside the orbit of Jupiter the sun's radiation converts some of its icy material into the gas that forms its characteristic halo and tail; the tail is made up of the particles streaming away from the sun under the pressure of radiation and the solar wind.

Some comets pass so close to the sun that they lose mass catastrophically and break into fragments. Others may make several passages, losing some material with each approach until they "die," probably becoming a residue of stony material that can no longer evolve gas. Still other comets may come close enough to one of the planets to have their orbit perturbed further until they end up behaving like one of the Apollo asteroids. Indeed, some investigators believe certain of the Apollo asteroids are extinct comets. The vast majority of comets, however, remain in a reservoir far beyond the outer planets.

How can one make systematic sense of the diverse small bodies of the solar system? Some are large, some small, some icy, some rocky, some naked and some shrouded in an atmosphere. One generalization that can be made is that a common by-product of the formation of large bodies in the solar system seems to have been the formation of associated small-body systems. The solar system itself contains one large body, the sun, and a number of satellites: the inner, terrestrial planets, the outer Jovian planets and the irregular comet swarm. Three of the giant planets are themselves miniature solar systems. Jupiter, Saturn and Uranus suggest the same pattern: small inner satellites, large intermediate satellites and irregular outer satellites.

A further consensus is developing with respect to the many interesting differences in composition that are evident among the planets and satellites of the solar system. Whereas certain other factors play a part, most of the differences among the planets can now be explained in terms of the differences in temperature at greater or lesser distances from the sun and the differences in gravitational fields that are a function of the different masses of the nine bodies. A detailed theory of the differences in composition and density among satellites and asteroids, however, is still lacking.

A basic approach to the question is to consider the distribution of mass among the components of the solar system. At the high-mass end of the scale the distribution is statistically irregular: the large bodies are overabundant. The distribution becomes smoother, however, as the number of bodies increases, their mass decreases and their diameters fall into the range of tens of kilometers. In fact, the number of small bodies, $N$, proves to be related to the mass, $m$, by a power law: $N$ is proportional (approximately) to $m$ raised to the $-2/3$ power.

Now, this proportional relation of number to mass happens to be characteristic of solids that broke up as a result of collisions. Some 15 years ago Gerald S. Hawkins of Boston University established the relation with respect to meteorites and proposed that the collisional fragmentation of asteroids was the mechanism responsible for it. In the mid-1960's Shoemaker pointed out that the debris blasted out of lunar craters showed a similar relation, although the exponent in the power law was closer to

1 than to $-2/3$. I have reviewed the geological literature and have also conducted experiments aimed at assessing the relation. I found that it existed in a wide range of collisional and grinding processes in nature. As the explosive energy (or the grinding pressure) is increased, the exponent changes from about $-2/3$ to about $-1$ or $-1.2$. The implication is that, at least as far as bodies with the dimensions of today's asteroids are concerned, the small-body mass distribution is a record of collisions. (Larger bodies, as they grew and swept up small bodies, established a different mass distribution.)

The craters that pock the moon, Mercury, Mars, Venus (as detected by radar probes) and the earth (as revealed by aerial and orbital photography) are, of course, direct evidence of the countless collisions that are characteristic of the solar system. These targets have been struck repeatedly by bodies ranging in size from less than a millimeter to more than 100 kilometers. The ages of lunar and terrestrial rocks show that the bombardment has continued for billions of years. When one counts and measures the craters, thereby reconstructing a record of the number and mass of the planetesimals responsible for them, the proportional relation between the two obeys about the same power law. Meteorites too show unmistakable signs of collisions between parent bodies. With these observations we can see some of the small-body history of the solar system beginning to emerge.

In the light of these collisional facts, combined with current theories of planet formation, it is informative to trace the evolutionary history of the small-body populations in three hypothetical stellar systems. The starting point should be when in each system a newly formed sun is surrounded by a hot but cooling nebula. In what we shall call System A the number of dust grains that initially condense in the nebula is relatively small; in System C the number is relatively large. System B, where the number is intermediate, is representative of the solar system. Initially the newly condensed planetesimals are roughly micron-size and their chemistry is determined by the chemistry and temperature of the cooling nebula.

The next evolutionary stage is based on a model of planetesimal accumulation proposed recently by Peter Goldreich of Cal Tech and William R. Ward of Harvard. Three quite different bell curves emerge. In System A, where the initial number of planetesimals was the

smallest, the micron-size particles have joined together under the force of mutual gravitational attraction until their average diameter has increased to perhaps a kilometer. In System B, with an intermediate number of initial planetesimals, the process of mutual attraction has produced about the same number of enlarged bodies, but their average diameter is substantially greater than one kilometer. In System C the planetesimals are not only more numerous but also larger than they are in the other two systems. The future histories of the three hypothetical systems will depend hereafter on the rate (and the violence) of the collisions between the enlarged bodies.

Before outlining the next steps it is worth noting that there is agreement between the hypothetical processes and the actual astrophysical observations. Frank J. Low of the University of Arizona and other observers have found young stars surrounded by infrared-radiating dusty nebulas that closely resemble the condensation models presented here. The dust or smoke, at a temperature of only a few hundred degrees Kelvin, obscures or partially obscures the central stars. Other young stars that can be dimly seen shining through their nebulas may represent systems where the dust grains have coalesced into asteroidlike bodies. Such star-and-nebula combinations may well be planetary systems in the making.

What happens next in the three hypothetical systems? The asteroidlike bodies sometimes collide with each other. Close approaches alter orbits, and particles with different relative velocities soon begin to interact. In the high-velocity collisions that follow, fragmentation begins to compete with the process of continued growth. The bell-shaped curve representing size variations is skewed because a succession of pulverizing collisions gives rise to large quantities of small-scale debris. The distribution of mass now reflects the power law applicable to fragmented masses.

The collision products steadily decrease in size as they collide and fragment. The submillimeter-size soil particles that accumulated on the moon during repeated cratering are examples. Much of this fine dust is probably swept out of each of the three systems, ejected into interstellar space by the pressure of radiation or the stellar wind. Larger, centimeter-size fragments may spiral in toward the central star, propelled by the Poynting-Robertson effect. At this point another connection between hypothesis and observation appears. For many years the origin of interstellar dust or smoke, which reveals itself by reddening the light of distant stars, has been a puzzle. The particles could not have coalesced in interstellar space; not enough atoms are present to have formed them. A few years ago George H. Herbig of the Lick Observatory and I independently proposed that the dust was protoplanetary matter entering interstellar space from the nebulas surrounding newly formed stars. Two processes might be involved. First, some of the original nebular dust would be expelled into space under the pressure of radiation and the stellar wind. Later on colliding planetesimals would provide a continuing supply of dust for expulsion into space.

Returning to the hypothetical stellar systems, the planetesimals could be fragmented to the point of literal disappearance if collision velocities are high. By the same token, if a few planetesimals grow to a substantial size during the initial period of low-velocity collisions, they can now continue to grow. The larger the body, the less likely it is that any subsequent collision would shatter it or knock off much of its mass. Moreover, the larger the body, the greater would be its gravitational field and its ability to capture passing particles. Thus once the growing bodies were past a certain threshold size they would have rapidly grown to planetary dimensions. That is why a mass-distribution diagram of our solar system shows an overabundance of mass allocated among a few large bodies.

In planetesimal-poor System A none of the accumulations become large enough even to approach the planetary threshold. Fragmentation overbalances accumulation until finally the only plan-etesimals in orbit around the star are in the subcentimeter to subkilometer range. This planetary system contains only asteroids. In System C the supply of planetesimals is large enough for bodies growing in all parts of the system to pass the planetary threshold. In the end the process of accumulation sweeps up almost all the planetesimals of subcentimeter to subkilometer size. Systems such as this might contain no asteroids at all; they might also grade into systems with two or more suns. About half of all stars are thought to belong to such systems.

Our solar system resembles the hypothetical System B. Why are some of our leftover planetesimals asteroids and others comets? It seems likely that the perturbing effects of massive Jupiter are responsible. Years ago Ernst Öpik demonstrated that Jupiter's perturbations could scatter and eject nearby planetesimals into the far reaches of the outer solar system. If the ejection took place early in the evolution of the solar system, when there was an abundance of small icy bodies in the cold outer solar system, then Jupiter's scattering effect would have created a deep-frozen reservoir of planetesimals: the comets. Thus both System B and System C could be expected to contain comets. William M. Kaula, George W. Wetherill and S. J. Weidenschilling have independently analyzed a second effect of Jupiter's perturbations. Not only were planetesimals scattered outward but also many would have been scattered into the asteroid belt and the inner solar system in highly energetic orbits. Interaction with these planetesimals may have prevented the evolution of the rocky asteroid population into a single moon-size body in orbit between Mars and Jupiter.

Other planetesimals became neither interstellar dust nor comets or asteroids but were scattered into elliptical orbits by interaction with nearby planets and finally collided with some planet or one of the planetary satellites. The largest of these collisions may have been key events in the establishment of present satellite and ring-system configurations. The smaller ones made craters.

Because we can compare crater densities on the moon with the ages of different moon rocks, the lunar cratering sequence during the past four billion years can be reconstructed in some detail. When the oldest-known moon rocks were formed, about four billion years ago, the rate of cratering was roughly 1,000 times higher than it is today. Half a billion years later, at the time when lava flowed over large areas of the moon, the supply of planetesimals (and hence the rate of cratering) had diminished to near its present level.

These interpretations are also of significance to the geology of the earth. The earliest cratering scars are no longer visible on the earth's eroded and churned surface but not all the terrestrial impact record has been destroyed. Michael R. Dence and other workers in Canada have identified a number of fossil craters on the Canadian shield, where the bedrock is roughly a billion years old. Taking into account the difference between the gravitational field of the earth and that of the moon, the rate of recent terrestrial cratering is found to be roughly consistent with the lunar rate. Thus the actions of some of the smaller bodies in the solar system have enabled us to write a few words on some heretofore blank pages of earth history.

12

INTERPLANETARY
PARTICLES AND FIELDS

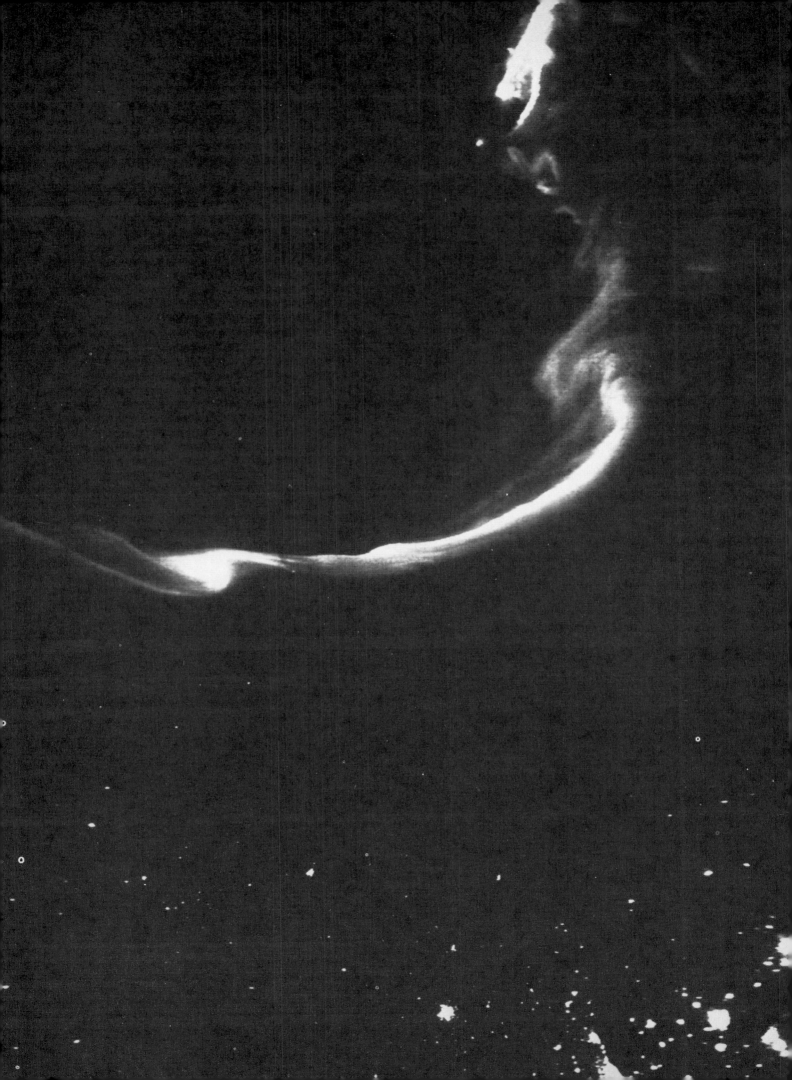

# Interplanetary Particles and Fields

JAMES A. VAN ALLEN

*A "wind" of charged particles blows out from the sun, punctuated by energetic bursts. These particles interact with the magnetic fields of the planets in intricate ways*

The medium through which the earth moves in its orbit around the sun contains on the average 10 particles of matter per cubic centimeter: five positive ions (mostly protons) and five electrons. In comparison there are some 27,000,000,000,000,000,000 molecules in a cubic centimeter of the earth's atmosphere at sea level. By terrestrial laboratory standards, then, the interplanetary medium is virtually a perfect vacuum. It might be thought that such a tenuous gas would have a negligible potential for producing physical phenomena. Yet a surprisingly rich variety of effects results from the interaction of the magnetic fields of the solar system and this "wind" of charged particles streaming outward from the sun. In the immediate vicinity of the earth, for example, solar-wind particles are captured by the earth's magnetic field and forced into a system of radiation belts that girdle the earth and are in turn enclosed within an elongated region called the magnetosphere. Such familiar geophysical phenomena as the polar auroras and the communications-disrupting ionospheric disturbances associated with "magnetic storms" owe their existence to the complex interplay between the solar wind and the earth's magnetosphere.

Since the advent of advanced radio-astronomical techniques and interplanetary spacecraft it has been found that other planets, most notably Jupiter, also have magnetospheres. The latest data in fact suggest that at least some of the energetic particles formerly classified as galactic cosmic rays are actually ions of the solar wind that have been accelerated to high energies in planetary magnetospheres.

Gases that are partly or fully ionized (in other words, gases in which some or all of the atoms are stripped of their electrons) are called plasmas. Since a plasma by definition consists of charged particles, it can conduct electricity and hence can sustain its own "entrained" magnetic field. In addition electrostatic fluctuations and other instabilities within a plasma can generate electromagnetic waves. Plasmas are sometimes regarded as constituting a fourth state of matter, since their physical properties differ so markedly from those of matter in the solid, liquid and neutral-gaseous states. The theoretical and experimental study of plasmas is one of the most active branches of contemporary physics. Major engineering efforts are under way to exploit plasmas for practical purposes, most notably in attempts to develop nuclear-fusion devices for the generation of electric power. The obstacles to building a successful fusion reactor are in fact primarily problems of plasma physics, not of nuclear physics.

Similarly, the role of plasmas in large-scale natural systems is at the forefront of space physics. Although the two disciplines deal with particles and fields on vastly different physical scales, there is a great deal in common in our understanding of them.

The concept of "corpuscular" radiation from the sun was invoked by various investigators in the late 19th and early 20th centuries to account for polar auroras and magnetic storms. At that time it was supposed that the corpuscles were energetic electrons emitted by active regions on the sun's surface. Later F. A. Lindemann suggested that the corpuscular radiation is a gas that consists of electrons and positive ions but is electrically neutral in bulk. The speed of these neutral "blobs" of gas was inferred from the observed time delay between a solar flare and the occurrence of a geomagnetic disturbance or an unusually bright auroral display. A delay of two days, for example, corresponds to an average speed of 870 kilometers per second over the mean distance of 150 million kilometers from the sun to the earth (or one astronomical unit).

During the period from 1900 to 1940 many workers, including Olaf K. Birkeland, Carl Störmer, Sydney Chapman, V. C. A. Ferraro, Julius Bartels and Hannes Alfvén, made noteworthy contributions to our understanding of the geophysical effects of the impact of a fast-moving plasma on the magnetic field of the earth. Alfvén in particular was the first to recognize the important role played by the magnetic field en-

AURORA BOREALIS, or "northern lights," seen from above in this extraordinary photograph made recently by a U.S. Air Force Weather Service satellite, is a particularly striking manifestation of the complex way in which matter from space interacts with the earth's magnetic field. The diaphanous auroral arc, an oval-shaped structure with characteristic eddylike folds every few hundred kilometers, covers much of central Canada in this view; the lights of Chicago and other cities of the northern U.S. can be seen near the bottom. It is now widely believed the fluorescent light of the polar auroras results from collisions between the atoms and molecules of the upper atmosphere and energetic charged particles from the sun that enter the atmosphere after being accelerated in the earth's magnetosphere. The incoming particles (mainly protons and electrons) are guided downward by the geomagnetic lines of force, which terminate in the earth's polar regions. The photograph was supplied by Ernest H. Rogers and David F. Nelson of the Aerospace Corporation.

trained in the plasma in the generation of auroral phenomena. The corpuscular streams came to be visualized as having their roots in active regions on the sun, with the bulk velocity of elements of the stream being radially outward from the center of the sun. By virtue of the sun's rotation, however, the geometric form of such streams in a nonrotating coordinate system would be an Archimedean spiral, resembling the form of a stream of water from a garden hose whose nozzle is rotated around an axis perpendicular to the outflowing stream [*see illustration below*]. In this picture a particular

stream would sweep past the earth at intervals of about 27 days, corresponding to the apparent rotational period of the sun as it is viewed from the earth. This feature provided a persuasive explanation for the 27-day period that had been observed for certain disturbances of the geomagnetic field.

It had been recognized before the 1950's that the tail of a comet contains dust and neutral (un-ionized) gas. Observational evidence now indicated that it contained ionized gas as well. This led Ludwig F. Biermann to suggest that the outflowing solar gas was a continuous,

although perhaps variable, feature of the interplanetary medium.

Then in 1958 an experiment I had designed for the first successful American earth satellite *Explorer 1* revealed the existence of large numbers of energetic electrons and protons trapped in a pair of radiation belts by the magnetic field of the earth. I attributed the existence of these particles to the entry of solar plasma, followed by an acceleration process of then unknown nature. The discovery of these remarkable geophysical features (later named Van Allen belts) led to the formation of a major new branch of

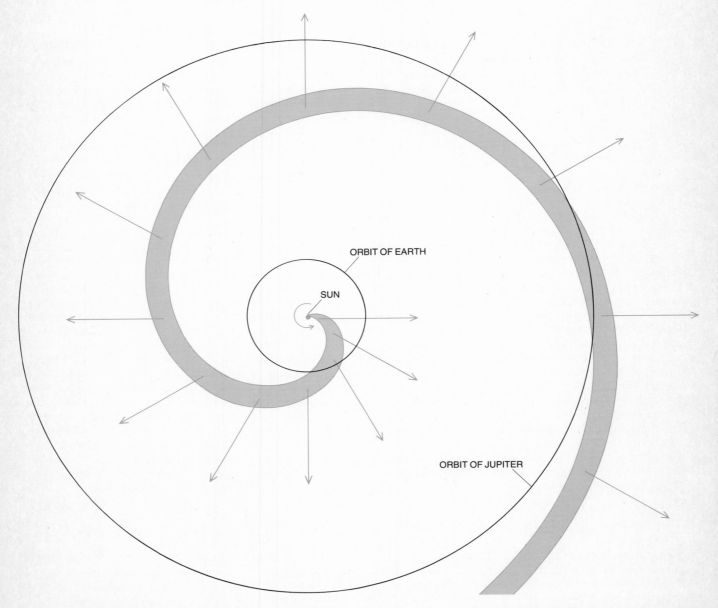

**GARDEN-HOSE MODEL** of the solar wind regards the outward stream of charged particles from the rotating sun as having the same geometric form as a stream of water from a rotating hose nozzle. In this idealized view an individual stream of particles emerging from an active region on the sun is represented by its projection on the sun's equatorial plane (as it would be seen by a stationary observer high above the sun's north pole). The boundaries of the stream in this perspective are Archimedean spirals. The velocity of

each element of the stream (*colored arrows*) is radially outward from the sun. For the purpose of the diagram this velocity was taken to be 350 kilometers per second and was assumed to be **independent** of time and distance from the sun. The "entrained" magnetic field set up by the stream is everywhere parallel to the spiral, with its overall direction being either outward or inward with respect to the sun. Model has been found to yield a good first-order approximation of observations that have been made by spacecraft.

space physics, now called magnetospheric physics. In the same year E. N. Parker published an important theoretical paper on the emission of hot ionized plasma from the solar corona. He named this outflowing gas the solar wind. He also put forward the idea that the magnetic field of the active solar region where the plasma had originated would be "frozen" into the plasma and would be drawn out into space in the average form of the Archimedean spiral. The magnetization of the ionized gas in this view can be visualized as an effect resulting from the persistent flow of a system of electric currents through the gas.

The first direct measurements of the solar wind were rather brief ones made in 1959 with a plasma detector aboard the Russian spacecraft *Luna 3;* two years later an American group made similar measurements with *Explorer 10.* The most extensive (and persuasive) of

the early observations of the solar wind were made in 1962 with an electrostatic analyzer carried by *Mariner 2* as it traversed interplanetary space en route to Venus. These observations confirmed the continuous flow of the solar wind and established its basic properties: It consisted largely of ionized hydrogen (in other words protons and electrons); its bulk flow was radially outward from the sun and was quite variable in velocity but generally in the range from 350 to 800 kilometers per second; it had an average density of 5.4 ions per cubic centimeter and an ion temperature of about 160,000 degrees Kelvin.

An earlier series of interplanetary magnetic-field measurements made by the space probe *Pioneer 2* were confirmed and greatly extended by the *Mariner 2* measurements. The average strength of the magnetic field was found to be about six gammas (six hundred-thousandths of a gauss); moreover, the

direction of the lines of force in the magnetic field, although quite variable, resembled the garden-hose model. Later work has emphasized the sectored structure of the interplanetary magnetic field, the discontinuous transitions from outward- to inward-directed magnetic-field vectors at the boundaries of such sectors and the persistence of a sectored structure through many rotations of the sun. The last feature provides a clear connection between interplanetary magnetic conditions and those localized in the sun's atmosphere, where there are usually only a few major sectors.

More detailed measurements of the solar wind have since been made by many workers, and several experiments are in progress. The most illuminating of the current series involves the two latest Pioneer spacecraft. *Pioneer 10* flew by Jupiter in December, 1973, and is on a hyperbolic escape trajectory from the solar system. *Pioneer 11* flew by Jupiter

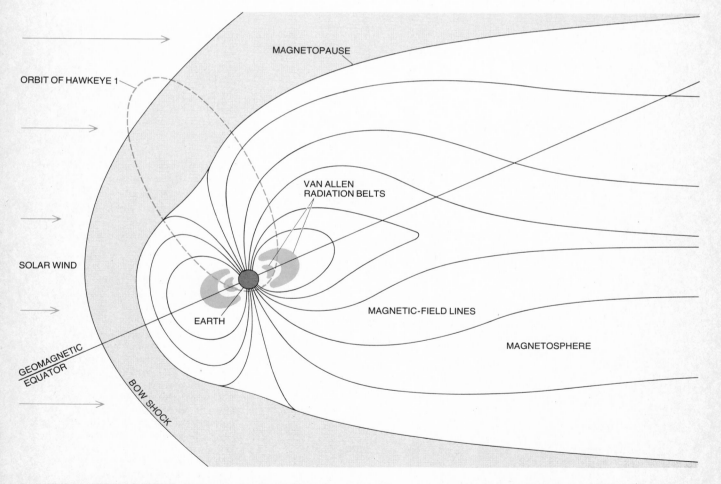

**EARTH'S MAGNETOSPHERE,** an elongated cavity hollowed out of the solar wind by the geomagnetic field, has been probed by a number of U.S. spacecraft, beginning with the early Explorer series, which in the late 1950's detected the comparatively nearby system of radiation belts that bear the author's name. Detailed observations of the magnetic field in the vicinity of the bow-shaped shock wave separating the magnetosphere from the solar wind are currently being conducted by the author's group at the University of Iowa, using a sensitive magnetometer carried aboard the artificial earth satellite *Hawkeye 1.* The orbit of this spacecraft (*broken colored ellipse*) is in a plane at right angles to the equatorial plane of the earth, with its apogee, or highest point, over the North Pole at a radial distance of 21 earth radii (more than 83,000 miles). Recent measurements of the direction of the interplanetary magnetic field outside the earth's magnetosphere, which are derived from the *Hawkeye 1* data, confirm the garden-hose model of the solar wind.

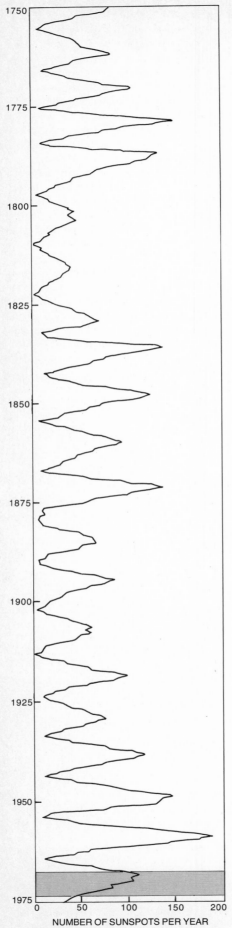

NUMBER OF SUNSPOTS PER YEAR

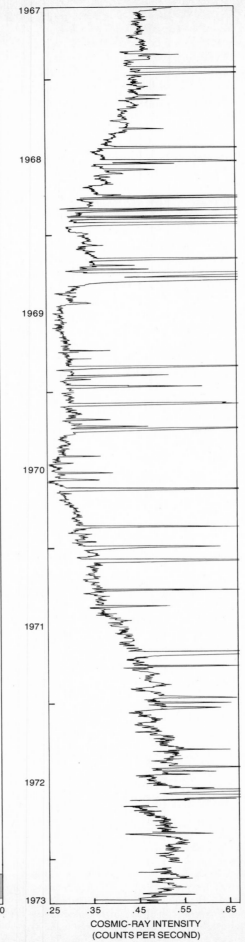

COSMIC-RAY INTENSITY
(COUNTS PER SECOND)

a year later and is now on its way to an encounter with Saturn in September, 1979 [*see illustration on page 131*]. The special interest of these interplanetary measurements lies in the extension of knowledge to very great distances from the sun. *Pioneer 10* is already more than eight astronomical units from the sun and is moving out of the solar system with an escape velocity of 2.4 astronomical units per year. It is a reasonable hope that data can be received from a radial distance as great as 20 astronomical units, a milestone that will be reached in October, 1979.

In spite of substantial theoretical conjecture to the contrary the properties of the solar wind have been found to change with distance from the sun in an essentially simple and continuous way. The average direction of the magnetic field continues to follow the garden-hose model. At five astronomical units the magnetic-field lines lie approximately in the central plane of the solar system but are nearly at right angles to a line connecting the observing spacecraft and the sun. (An alternative way of viewing the matter is to suppose that the magnitude of the radial component of the magnetic field has declined as the inverse square of the distance from the sun whereas the magnitude of the azimuthal component has declined as the inverse first power.) There continue to be fluctuations in magnetic-field strength similar to those observed in the vicinity of the earth's orbit out to a distance of at least five astronomical units.

Simultaneous measurements of the solar wind are being made by the Pioneer spacecraft using an elaborate plas-

COSMIC-RAY INTENSITY has been observed to vary over an 11-year period in approximate synchrony with the long-established 11-year sunspot cycle, with the maximum value of cosmic-ray intensity coming at the time of minimum solar activity and vice versa. This inverse relation, originally discovered as a result of ground-based observations, has now been confirmed by satellite-borne detectors. The first curve at left gives the variation in solar activity for the period from 1750 to 1975. The second curve shows the results of a six-year study by the author of the intensity of high-energy cosmic-ray protons in the interplanetary medium at a distance from the sun of about one astronomical unit. Detector used to make measurements is on lunar-orbiting spacecraft *Explorer 35*. Overall shape of curve represents variation of the cosmic rays; the numerous "spikes" are caused by energetic particles from the sun. Note that the cosmic-ray intensity has a minimum value in 1969–1970 near time of maximum solar activity.

ma analyzer that determines the bulk velocity, the ion temperature, the ionic composition and the density of the ions. The average velocity of the solar wind changes imperceptibly over the range from one astronomical unit to five astronomical units, but the range of fluctuations diminishes markedly. The latter feature appears to be caused by interactions between fast and slow ion streams, producing a tendency toward uniformity. The average ion temperature has dropped by about a factor of two and the average ion density has declined in approximate accord with the inverse square of the distance from the sun. The latter result corresponds to a spherically symmetrical radial flow, although the random fluctuations are so large that they preclude an accurate test of the inverse-square law.

Meanwhile the German-American spacecraft *Helios* has been sampling the solar wind and the interplanetary magnetic field inside the earth's orbit; this vehicle came closest to the sun in mid-March, 1975, at a point less than a third of an astronomical unit from the sun's center. An experiment of special interest on *Helios* is one aimed at measuring radio noise generated within the interplanetary medium itself. This noise, in the frequency range from 10 to 100,000 hertz (cycles per second), is a revealing measure of plasma instabilities and of the mechanisms by which energy is ex-

changed between protons and electrons in a hot plasma.

Every direct measurement of the solar wind so far has been made within less than nine degrees from the plane of the sun's equator, which is inclined to the ecliptic (the plane of the earth's orbit) by seven degrees. There are two reasons for this. First, the "slingshot" effect of the earth's orbital velocity (30 kilometers per second) is a central feature of the launch dynamics of any interplanetary spacecraft. In fact, for a circular orbit with a radius of one astronomical unit an orbital inclination to the ecliptic of 20 degrees is the practical upper limit of existing launch-vehicle capability. Second, measurements of the solar wind have been made in most cases by multipurpose spacecraft that are targeted for close planetary encounters, thus requiring orbits in planes only slightly inclined to the ecliptic. Important but less quantitative data on the properties of the solar wind at higher solar latitudes are being obtained by other methods, such as the observation of comet tails and the analysis of "scintillations" induced in radio signals from stellar sources by irregularities in the interplanetary plasma.

Ground-based observations of the sun's disk show that solar activity is not spherically symmetrical but rather depends strongly on solar latitude. Because

of this fact many investigators feel a compelling need for direct observations at the higher solar latitudes. An out-of-the-ecliptic mission can be achieved by a spacecraft that "flies by" Jupiter in such a way that its subsequent trajectory is inclined at 90 degrees to the ecliptic and passes over the pole of the sun at a distance of one astronomical unit. Another possibility is to gradually tilt the orbital plane of a spacecraft circling the sun at one astronomical unit to an inclination of about 40 degrees by the technique of solar electric propulsion. That technique utilizes sunlight collected by solar cells to generate electricity, which in turn is used to power an "ion gun" for auxiliary propulsion. The thrust resulting from the ejection of energetic ions would be directed at right angles to the path of the spacecraft.

An out-of-the-ecliptic mission would for the first time enable investigators to study the interplanetary medium in three dimensions. Such a mission would also make it possible to obtain good observations of the polar regions of the sun itself. Meanwhile the best data available will be those gathered by *Pioneer 11*, which will reach a maximum solar latitude of 20 degrees at a distance of about five astronomical units in early 1977 as it travels on an arc across the solar system en route to Saturn.

The outward flow of the solar wind is presumably arrested at some point by

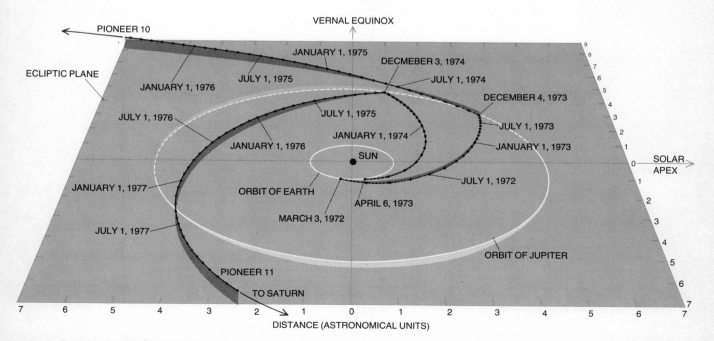

**TWO LATEST PIONEER MISSIONS** have extended the investigation of the solar wind to very great distances from the sun. This three-dimensional diagram shows the trajectories of *Pioneer 10* (launched on March 3, 1972) and *Pioneer 11* (launched on April 6, 1973) in relation to the ecliptic (the plane of the earth's orbit). The black dots represent the positions of the spacecraft at the beginning of each month with respect to a heliocentric coordinate system, with longitude measured counterclockwise from the vernal equinox. The discontinuities in the trajectories of the two spacecraft were caused by close encounters with the planet Jupiter on December 4, 1973 (*Pioneer 10*), and December 3, 1974 (*Pioneer 11*). The solar apex indicates the direction of the sun's motion.

CLUES TO PROPERTIES OF THE SOLAR WIND can be obtained from the study of comet tails, which have been known for some time to have two distinct components: one consisting of dust and neutral (un-ionized) gas and the other consisting of ionized gas. In most comet photographs the dust tail (which generally lags behind the ionized-gas tail) is the more prominent component. In this photograph of Comet Mrkos, which was made on August 23, 1957, with the 48-inch Schmidt telescope on Palomar Mountain, the two types of tail can be distinguished with unusual clarity. The highly structured tail of ionized gas, driven radially outward from the sun by the solar wind, precedes the diffuse, comparatively featureless dust tail across the sky. (The motion of the comet is toward the upper left in photograph.) Comet Mrkos had made its closest approach to the sun at a distance of .355 astronomical unit on August 1; at the time photograph was made the comet was outward bound at a point .76 astronomical unit from sun and 1.21 astronomical units from the earth.

its encounter with the interstellar gas, which is estimated to have a density of less than one atom per cubic centimeter in the general vicinity of the solar system. This hypothetical boundary has been called the heliopause. The velocity of the solar system with respect to neighboring stars is about 20 kilometers per second toward a point in the sky called the solar apex. On the assumption that the nearby interstellar gas is at rest in this reference system, the distance to the heliopause is presumably least in the direction of the solar apex because of the opposing velocity of the interstellar gas in that direction. Theoretical estimates of that distance have ranged anywhere from five to 300 astronomical units. Instruments on *Pioneer 10* have shown no evidence for such a transition boundary out to eight astronomical units; unfortunately the spacecraft is now receding from the sun in a direction nearly opposite the solar apex.

So much for the properties of the solar wind itself. What about its properties as a medium through which charged particles of much greater energies travel? Collisions between particles are for the most part of minor interest here, since the mean free path for such collisions is on the order of 100 astronomical units. Energetic charged particles interact with the interplanetary medium mainly through "kinks," or irregularities, in the entrained magnetic field. Any charged particle moving through a uniform magnetic field is forced into a helical orbit with the axis of the helix along the lines of force in the magnetic field. The radius of the helix is directly proportional to the momentum of the particle and inversely proportional to the strength of the magnetic field. A particle with an orbital radius substantially less than the radius of curvature of a kink in the local magnetic field moves along a helical path whose center of curvature, or guiding center, approximately follows the line of magnetic force. A particle with an orbital radius substantially greater than the kink's curvature has its direction of motion only slightly affected and moves in a path corresponding to the larger-scale features of the magnetic field. A particle with an orbital radius comparable to the curvature of the field, however, will often have its trajectory drastically changed. The latter process is called scattering by a magnetic irregularity. Recent measurements of the magnetic field in the interplanetary medium, in addition to showing the average tendency toward the garden-hose model, indicate many local irregularities in both the direction and the magnitude of the magnetic field.

In 1954 Scott E. Forbush published the results of a long series of ground-based observations showing that the intensity of the secondary products of galactic cosmic rays exhibits an 11-year periodicity that is approximately synchronized with the 11-year cycle of solar activity. The maximum value of cosmic-ray intensity comes at the time of minimum solar activity and vice versa. This effect can be explained as follows: Cosmic rays from distant sources approach the solar system from all directions more or less uniformly. As they enter the solar system they are scattered by magnetic irregularities in the interplanetary medium. By virtue of the outward radial motion of such irregularities the collisions give a small but generally outward component to the velocity of the scattered particle. This effect tends to diminish the intensity of the cosmic rays within the solar system, the magnitude of the effect being greater during periods of maximum solar activity and hence of maximum average solar-wind velocity and magnetic turbulence. The total intensity of the galactic cosmic radiation at the earth's orbit is roughly twice as great during a "solar minimum" as it is during a "solar maximum." This modulation factor is greatest at the lower energies and is negligible at energies above $10^{11}$ electron volts. Presumably the intensity of galactic cosmic rays in

the vicinity of the solar system but far enough from the sun to be unaffected by the interplanetary magnetic field remains constant over periods of perhaps millions of years, and certainly over periods far greater than 11 years. Therefore a necessary corollary of the observed temporal variation of galactic cosmic rays must be a radial gradient of intensity directed outward from the sun.

In principle both effects should be calculable from the observed incidence of magnetic irregularities and from an assumed form of the energy spectrum of the galactic cosmic radiation in interstellar space. Theoretical efforts to solve this difficult convection-diffusion problem have been illuminating but so far largely unsuccessful. The direct measurements of total cosmic-ray intensity with *Pioneer 10* and *Pioneer 11* show an essentially zero gradient between one astronomical unit and eight astronomical units from the sun, whereas the prevailing theory predicts an increase on the order of 60 percent over this range. These results suggest that the heliopause is much more remote from the sun than eight astronomical units.

Another major class of energetic particles in the solar system comprises the particles emitted in solar flares. The first clear example of this phenomenon was observed by Forbush and others in 1942 with ground-level ionization chambers. For a number of years thereafter only a few further cases were observed.

High-altitude balloons, rockets, earth satellites and interplanetary spacecraft have now enormously expanded observational knowledge of these energetic solar particles. The energy range of the particles (which at one time were inappropriately called solar cosmic rays) extends from a few thousand electron volts to hundreds of millions of electron volts.

Among the particles that have been identified so far are electrons, protons and the nuclei of helium, carbon, nitrogen, oxygen and heavier atoms, all of which are present in the sun's atmosphere. Protons and electrons are the dominant species, as might be expected considering that hydrogen is the most abundant element in the sun's atmosphere and electrons are a universal constituent of matter. These particles are accelerated in the early "flash" phase of solar flares, escape from the sun and diffuse outward through the interplanetary magnetic field. Electrons, by virtue of their higher velocity, usually exhibit a fairly smooth profile of intensity with respect to time, as one might expect for diffusion through a static medium of scattering centers. The same is true of protons with energies on the order of 100 million electron volts and more. Lower-energy protons and heavier ions, however, exhibit very complex intensity-time profiles, and "spikes" of proton intensity are often associated with shock waves and other discontinuities in the interplanetary medium.

This qualitative difference is made in-

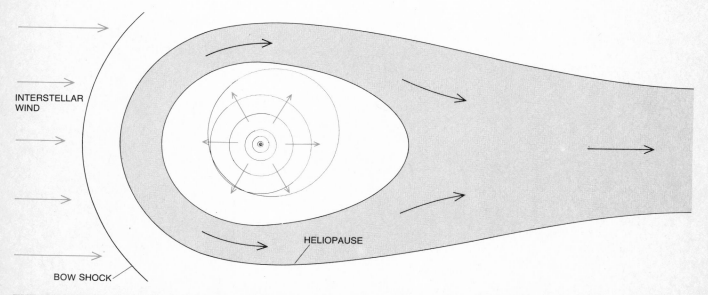

**THE HELIOSPHERE,** an elongated region resembling the earth's magnetosphere, is believed to enclose the entire solar system on its journey through interstellar space. (The solar system moves with respect to the neighboring stars at a velocity of about 20 kilometers per second toward the solar apex.) The hypothetical boundary where the outward flow of the solar wind is arrested by its encounter with the much more tenuous interstellar medium is called the heliopause. Because of the opposing (relative) velocity of the interstellar gas from the direction of the solar apex the distance to the heliopause is presumably least in that direction. Theoretical estimates of this distance have ranged from five to 300 astronomical units; to date *Pioneer 10* has found no evidence of such a transition out to eight astronomical units. Here heliopause is arbitrarily drawn at 75 astronomical units. Black ellipses are planetary orbits.

telligible by noting that the velocity of a proton with a kinetic energy of 300,000 electron volts (a typical value) is only 20 times greater than the solar-wind velocity. In contrast, the velocity of a typical electron with a kinetic energy of 30,000 electron volts is more than 250 times greater than the solar-wind velocity. The result is that the motion of the lower-energy ions is strongly influenced by convective effects caused by the outward motion of the magnetic scattering centers. In addition the energies of these particles are changed by encounters with interplanetary shock waves and other moving discontinuities.

Hundreds of distinct energetic-solar-particle events have been observed in great detail by spacecraft-borne instruments. Many events have been observed by two or more spacecraft at widely separated points. Each event is different in detailed characteristics. Observations are commonly made of energy spectra, the abundance of electrons, the abundances of nuclei of various elements and the distribution of each constituent with respect to angle and intensity over a period of time. The gross features of the intensity-time curves for electrons are roughly understood in terms of diffusion in the interplanetary magnetic field, but the intensity-time curves for ions (particularly low-energy ions) are much more complex and reflect convective-transport effects and local acceleration processes as well as diffusional effects. At times the convective processes are dominant enough to establish a positive radial gradient of intensity late in an event; diffusive flow is then inward toward the sun rather than outward from the original source, as is the case early in the event. Particle intensities as

great as half a million protons per square centimeter per second have been observed at energies greater than 300,-000 electron volts. Particularly intense events last for several days and may present a significant radiation hazard to long manned space missions. Most of these energetic-particle events are conclusively identified with observed solar flares. Those that are not can reasonably be attributed to "back side" flares on the unobserved hemisphere of the sun.

The process by which particles are accelerated in solar flares is thought to be associated with the collapse of a magnetic field at the site of the flare. The relative intensities of the various ionic species have a rough resemblance to the relative abundances of corresponding atoms in the sun's atmosphere as determined by traditional spectroscopic methods. It has been suggested that by studying the relative abundances of the various species of energetic solar ions one might derive a good estimate of the relative abundances of elements in the sun. Unfortunately the observed relative ion intensities vary with energy, vary from event to event and vary with time during individual events. It appears that differential effects occur during the acceleration process and during the interplanetary propagation. Such variations are more pronounced at the lower energies.

From all the above it is clear that the sun is the major performer in producing phenomena involving particles and fields in interplanetary space. Nonetheless, the magnetospheres of the earth and Jupiter are microcosms of such phenomena. Moreover, in these microcosms both large-scale and fluctuating electric

fields also play a central role. The magnetospheres of both the earth and Jupiter are now known to be weak emitters of energetic particles.

If the heliosphere is as large as several hundred astronomical units, it is not altogether unreasonable to consider the possibility, as Alfvén has long done, that a significant fraction of the "galactic" cosmic radiation (particles with an energy of less than, say, $10^{11}$ electron volts) originates within the solar system. This possibility, however, is contrary to the prevailing view, which attributes the origin of all cosmic rays to remote stellar and interstellar processes, such as those that are thought to take place in supernovas.

The interplanetary medium is traversed by all kinds of electromagnetic waves, originating in the sun, the planets and sources outside the solar system. Electromagnetic waves whose frequency is less than the local plasma frequency in the interplanetary medium are absorbed near the source and cannot propagate. This cutoff frequency is proportional to the square root of the local electron density. At a distance of one astronomical unit from the sun, for example, the cutoff frequency is 20,000 hertz; as a result no electromagnetic wave of lower frequency can reach the vicinity of the earth. Above that frequency the propagation of electromagnetic waves is affected in various ways, but the influence of the interplanetary medium progressively diminishes with increasing frequency and is usually negligible at frequencies greater than 100 million hertz. It is this relative transparency of the interplanetary medium to electromagnetic waves that makes possible the subject of astronomy.

# THE AUTHORS
# BIBLIOGRAPHIES
# INDEX

# THE AUTHORS

CARL SAGAN ("The Solar System") is professor of astronomy and space sciences at Cornell University, where he also directs the Laboratory for Planetary Studies. He is a graduate of the University of Chicago and received his Ph.D. there in 1960. Before joining the Cornell faculty in 1968 he taught at Harvard University and worked at the Smithsonian Astrophysical Observatory in Cambridge, Mass. An expert on the physics of planetary atmospheres, planetary surface conditions and the possibility of extraterrestrial life, Sagan has served as a consultant and experimenter on a number of U.S. planetary missions, including the forthcoming Viking landing on Mars. He is editor-in-chief of *Icarus: The International Journal of Solar System Studies,* and currently serves as chairman of the division for planetary sciences of the American Astronomical Society.

A. G. W. CAMERON ("The Origin and Evolution of the Solar System") is professor of astronomy at Harvard University and associate director for planetary sciences of the Center for Astrophysics in Cambridge, Mass. A native of Winnipeg, he obtained his Ph.D. in experimental nuclear physics from the University of Saskatchewan in 1952. His early experience in nuclear physics, he writes, "led me into astrophysics, since I became involved in unraveling the puzzle of the origin of the elements by stellar nucleosynthesis shortly after obtaining my Ph.D. Before one can understand quantitatively the origin of the elements, however, one must know the elemental abundances and hence understand the best sources of those abundances in the solar system: the meteorites. One cannot understand the meteorites, however, without understanding the origin

of the solar system in general. So there you see the tortuous path by which I arrived at my interest in the subject of this article." Cameron was a senior scientist at the National Aeronautics and Space Administration's Goddard Institute for Space Studies in New York City and professor of space physics at Yeshiva University before taking up his present posts in 1973.

E. N. PARKER ("The Sun") is Distinguished Service Professor in the departments of physics, astronomy and astrophysics at the University of Chicago. A graduate of Michigan State University, he acquired his Ph.D. from the California Institute of Technology in 1951. He taught mathematics and physics at the University of Utah until 1955, when he moved to Chicago to become a research associate of the Institute for Nuclear Studies. Parker is perhaps best known for his development of the idea of the solar wind. His current research interests, he reports, are focused "on the general role of magnetic fields in creating activity in the sun, in other stars and in the galaxy."

BRUCE C. MURRAY ("Mercury") is professor of planetary science at the California Institute of Technology; on his return from a sabbatical leave early in 1976 he will become the next director of the Jet Propulsion Laboratory, which is operated by Cal Tech for the National Aeronautics and Space Administration. A geologist by training, he received his Ph.D. from the Massachusetts Institute of Technology in 1955. Since going to Cal Tech in 1960 he has participated in five of the Mariner missions, most recently as head of the group responsible for the *Mariner 10* television experiment, which produced the first closeup

pictures of Mercury and Venus. In June he served as chairman of the First International Colloquium on Mercury, which was held at Cal Tech.

ANDREW and LOUISE YOUNG ("Venus") do research at Texas A&M University. Andrew Young studied physics as an undergraduate at Oberlin College and astronomy as a graduate student at Harvard University, obtaining a Ph.D. from the latter institution in 1962. He taught astronomy at Harvard and at the University of Texas for several years and later worked at the Aerospace Corporation and the Jet Propulsion Laboratory before joining the faculty of Texas A&M in 1973. Louise Young was graduated with highest honors in engineering from the University of California at Los Angeles in 1958. She acquired her Ph.D. from the California Institute of Technology in 1963. Before going to Texas A&M she taught thermodynamics and statistical mechanics at U.C.L.A., worked as a consultant in the aircraft industry and did research at Cal Tech, the Jet Propulsion Laboratory and the University of Texas. The Youngs write: "We each trickled into the field of planetary atmospheres from a very different direction: A. T. Y. from an interest in photoelectric photometry, which led to the photometric study of atmospheres and their optical properties, and L. G. Y. from a background in radiative transfer and rocket exhaust, which like the atmospheres of terrestrial planets contains water and carbon dioxide."

RAYMOND SIEVER ("The Earth") is professor of geology at Harvard University, where he heads the Committee on Experimental Geology and Geophysics. His academic degrees are all from the University of Chicago: a B.S. in

1943, an M.S. in 1947 and a Ph.D. in 1950. Beginning as a research assistant in 1943, Siever was associated for many years with the Illinois Geological Survey. He came to Harvard in 1956 on a National Science Foundation postdoctoral fellowship and stayed on to join the faculty a year later. Among his current projects, he reports, are studies of "the coal resources of underdeveloped countries, the origin of sulfur in high-sulfur coals and the prospects for recovery of large deposits of oil and gas at the edges of the continents under the sea.... I am also in process of trying to reevaluate the history of the Atlantic Ocean and the early continents at the time of coal-bed formation in the late Paleozoic." Siever is the coauthor (with Frank Press) of the introductory college-geology textbook *Earth,* published recently by W. H. Freeman and Company.

JOHN A. WOOD ("The Moon") is a member of the staff of the Smithsonian Astrophysical Observatory of the Center for Astrophysics in Cambridge, Mass., where he has devoted much of the past few years to the study of rocks from the moon. "I began teaching courses in geology at Harvard two years ago," he writes, "but apart from that I have worked full time on lunar research since the beginning of the Apollo program. From 1971 to 1973 I was vice-chairman of the Lunar Sample Analysis Planning Team, the advisory group that recommends to NASA on allocations of lunar samples to investigators in the program. My principal other interest (or ambition) continues to be the study of meteorites. I am eager to resume work on these enigmatic objects, which appear to be witnesses to the origin of the solar system and perhaps even earlier events." Wood is a graduate of the Virginia Polytechnic Institute and holds a Ph.D. in geology from the Massachusetts Institute of Technology.

JAMES B. POLLACK ("Mars") is a theoretical astrophysicist on the staff of the Ames Research Center of the National Aeronautics and Space Administration. He was educated at Princeton University (A.B. in physics, 1960), the University of California at Berkeley (M.A. in physics, 1962) and Harvard University (Ph.D. in astronomy, 1965). He joined the research staff at Ames in 1970 after working at the Smithsonian Astrophysical Observatory in Cambridge, Mass., and the Center for Radiophysics and Space Research of Cornell University. Pollack has participated in several unmanned-spacecraft missions to the planets, including the 1971 *Mariner 9* orbiter mission to Mars, for which he headed the group responsible for obtaining television pictures of the Martian satellites. He is currently working on experiments for the 1976 Viking missions to Mars and the 1978 Pioneer mission to Venus. In addition to his involvement in such missions, he has spent a good part of the past few years investigating the causes of major climatic changes on the earth. "In so doing," he remarks, "I have applied some of the computational techniques I have developed for other planets to our own planet."

JOHN H. WOLFE ("Jupiter") is chief of the space-physics branch of the space-science division of the National Aeronautics and Space Administration's Ames Research Center, which he joined soon after receiving his Ph.D. from the University of Illinois in 1960. Since then his main concern has been the measurement of the interplanetary solar wind and its interactions with the planets. He has been the principal investigator for the solar-wind experiments on a total of 12 unmanned space probes. In addition he is chief scientist for the Pioneer missions to Jupiter and as such is responsible for their overall scientific planning and coordination.

DONALD M. HUNTEN ("The Outer Planets") is a physicist with the planetary sciences division of the Kitt Peak National Observatory. A native of Montreal, he was educated at the University of Western Ontario and McGill University, receiving his Ph.D. in physics from McGill in 1950. He spent the next 13 years at the University of Saskatchewan, "studying the aurora and the upper atmosphere and teaching physics and astronomy." In 1963 he joined a group assembled at Kitt Peak to investigate planetary atmospheres. He comments: "Currently my time is divided among all the planets, including the earth, where I have been involved in some of the efforts to assess threats to atmospheric ozone. My chief recreation is playing early music with the Collegium Musicum of the University of Arizona; my instruments are bassoon, flute, recorders and krummhorns. I also maintain my wife's harpsichord and dabble in harpsichord building."

WILLIAM K. HARTMANN ("The Smaller Bodies of the Solar System") works in Tucson for the Planetary Science Institute, a division of Science Applications, Inc. He studied physics as an undergraduate at Pennsylvania State University, then went on to obtain an M.S. in geology and a Ph.D. in astronomy from the University of Arizona. Since the early 1960's he has worked primarily on the origin and evolution of planets and planetary systems. His 1972 textbook, *Moons and Planets,* has been widely used as an introduction to planetary science. He was a coinvestigator on the *Mariner 9* mission, and in 1974 he coauthored (with Odell Raper) a copiously illustrated NASA publication, *The New Mars,* describing the *Mariner 9* results.

JAMES A. VAN ALLEN ("Interplanetary Particles and Fields") is Carver Professor of Physics and head of the department of physics and astronomy at the University of Iowa. He did his undergraduate work at Iowa Wesleyan College and his graduate work at the University of Iowa, acquiring his Ph.D. in physics in 1939. He returned to Iowa in 1951 after devoting several years to research elsewhere, including the Carnegie Institution of Washington and Johns Hopkins University. Van Allen is noted as a pioneer in the use of balloons, rockets and spacecraft for scientific measurements; in the late 1950's he discovered the radiation belts of the earth with apparatus on the first successful U.S. satellite, *Explorer 1,* and confirmed the discovery soon thereafter with *Explorer 3.* He has since participated in some 20 satellite and planetary missions, doing experiments on the behavior of particles and fields in the earth's magnetosphere, in interplanetary space and in the vicinity of Venus, Mars and Jupiter. One of his main research interests at the moment is in measuring the polar magnetic field of the earth at high altitudes, using *Hawkeye 1,* the seventh satellite to be built under his supervision at the Iowa campus. During the past two years he has also helped develop plans for a future Mariner mission to Jupiter and Uranus and a future Pioneer orbiter mission to Jupiter.

# BIBLIOGRAPHIES

*Readers interested in further reading on the subjects covered by articles in this issue may find the list below helpful.*

## THE SOLAR SYSTEM

THE NEW MARS. Edited by William K. Hartmann and Odell Raper. NASA SP-337. U.S. Government Printing Office, 1974.

PIONEER ODYSSEY: ENCOUNTER WITH A GIANT. Prepared by Richard O. Fimmel, William Swindell and Eric Burgess. NASA SP-349. U.S. Government Printing Office, 1974.

THE COSMIC CONNECTION: AN EXTRATERRESTRIAL PERSPECTIVE. Carl Sagan. Dell Publishing Co., 1975.

## THE ORIGIN AND EVOLUTION OF THE SOLAR SYSTEM

ORIGIN OF THE SOLAR SYSTEM. Edited by Robert Jastrow and A. G. W. Cameron. Academic Press, 1963.

EVOLUTION OF THE PROTOPLANETARY CLOUD AND FORMATION OF THE EARTH AND PLANETS. V. S. Safronov. Israel Program for Scientific Translations, Jerusalem, 1972.

SYMPOSIUM ON THE ORIGIN OF THE SOLAR SYSTEM. Edited by Hubert Reeves. Edition du Centre National de la Recherche Scientifique, Paris, 1972.

NUMERICAL MODELS OF THE PRIMITIVE SOLAR NEBULA. A. G. W. Cameron and M. R. Pine in *Icarus*, Vol. 18, No. 3, pages 377–406; March, 1973.

EARLY CHEMICAL HISTORY OF THE SOLAR SYSTEM. Lawrence Grossman and John W. Larimer in *Reviews of Geophysics and Space Physics*, Vol. 12, No. 1, pages 71–101; February, 1974.

## THE SUN

OUR SUN. Donald H. Menzel. Harvard University Press, 1959.

THE SUN. Karl O. Kiepenheuer. University of Michigan Press, 1959.

A STAR CALLED THE SUN. George Gamow. Viking Press, 1964.

EARLY SOLAR PHYSICS. A. J. Meadows. Pergamon Press, 1970.

## MERCURY

SCIENCE, Vol. 195; July 12, 1974.

MAGNETIC FIELD OF MERCURY CONFIRMED. N. F. Ness, K. W. Behannon, R. P. Lepping and Y. C. Whang in *Nature*, Vol. 255, pages 204–205; May 15, 1975.

JOURNAL OF GEOPHYSICAL RESEARCH, Vol. 80, No. 17; June 10, 1975.

FLIGHT TO MERCURY. Bruce C. Murray and Eric Burgess. Columbia University Press, in press.

## VENUS

THE PLANET VENUS. Patrick Moore. Macmillan, Inc., 1961.

THE PLANET VENUS. Carl Sagan in *Science*, Vol. 133, No. 3456, pages 849–858; March 24, 1961.

COMMENT ON "THE COMPOSITION OF THE VENUS CLOUD TOPS IN LIGHT OF RECENT SPECTROSCOPIC DATA." L. D. G. Young and A. T. Young in *The Astrophysical Journal*, Vol. 179, No. 1, Part 2, pages L39-L43; January 1, 1973.

INFRARED SPECTRA OF VENUS. L. G. Young in *International Astronomical Union Symposium, Vol. 65: Exploration of Planetary Systems—Part II, Terrestrial Planets*, edited by A. Woszczyk and C. Iwaniszewska. D. Reidel, 1974.

JOURNAL OF THE ATMOSPHERIC SCIENCES, Vol. 32, No. 6; June, 1975.

## THE EARTH

SPECULATIONS ON THE EARTH'S THERMAL HISTORY. Francis Birch in *The Geological Society of America Bulletin*, Vol. 76, No. 2, pages 133–153; February, 1965.

HISTORY OF THE EARTH: AN INTRODUCTION TO HISTORICAL GEOLOGY. Bernhard Kummel. W. H. Freeman and Company, 1970.

EVOLUTION OF THE EARTH. Robert H. Dott, Jr., and Roger L. Batten. McGraw-Hill Book Company, 1971.

MODEL FOR THE EARLY HISTORY OF THE EARTH. S. P. Clark, Jr., K. K. Turekian and L. Grossman in *The Nature of the Solid Earth*, edited by Eugene C. Robertson. McGraw-Hill Book Company, 1972.

SEDIMENTARY CYCLING IN RELATION TO THE HISTORY OF THE CONTINENTS AND OCEANS. R. M. Garrels, F. T. MacKenzie and R. Siever in *The Nature of the Solid Earth*, edited by Eugene C. Robertson. McGraw-Hill Book Company, 1972.

FORMATION OF THE EARTH'S CORE. Don L. Anderson and Thomas C. Hanks in *Nature*, Vol. 237, No. 5355, pages 387–388; June 16, 1972.

EARTH. Frank Press and Raymond Siever. W. H. Freeman and Company, 1974.

COMPARISON OF EARTH AND MARS AS DIFFERENTIATED PLANETS. Raymond Siever in *Icarus*, Vol. 22, No. 3, pages 312–324; July, 1974.

## THE MOON

GEOLOGY OF THE MOON: A STRATIGRAPHIC VIEW. Thomas A. Mutch. Princeton University Press, 1970.

LUNAR BASIN FORMATION AND HIGHLAND STRATIGRAPHY. K. A. Howard, D. E. Wilhelms and D. H. Scott in *Reviews of Geophysics and Space Physics*, Vol. 12, No. 3, pages 309–327; August, 1974.

STRUCTURE OF THE MOON. M. Nafi Toksöz, Anton M. Dainty, Sean C. Solomon and Kenneth R. Anderson in *Reviews of Geophysics and Space Physics*, Vol. 12, No. 4, pages 539–567; November, 1974.

LUNAR SCIENCE: A POST-APOLLO VIEW.

Stuart Ross Taylor. Pergamon Press, 1975.

## MARS

MARS AS VIEWED BY MARINER 9. NASA SP-329. U.S. Government Printing Office, 1974.

## JUPITER

HANDBOOK OF THE PHYSICAL PROPERTIES OF THE PLANET JUPITER. Edited by C. M. Michaux. NASA SP-3031. U.S. Government Printing Office, 1967.

SPACE VEHICLE DESIGN CRITERIA (ENVIRONMENT): THE PLANET JUPITER [1970]. N. Divine. NASA SP-8069. National Aeronautics and Space Administration, December, 1971.

SCIENCE, Vol. 183, No. 4122, pages 301–324; January 25, 1974.

PIONEER 10 MISSION: JUPITER ENCOUNTER in Journal of Geophysical Research, Vol. 79, No. 25, pages 3487–3886; September 1, 1974.

SCIENCE, Vol. 188, No. 4187, pages 445–477; May 2, 1975.

## THE OUTER PLANETS

THE PLANET SATURN. A. F. O'Donel Alexander. Faber and Faber, Ltd., London, 1962.

THE PLANET URANUS. A. F. O'Donel Alexander. Faber and Faber, Ltd., London, 1965.

THE ATMOSPHERE OF TITAN. Edited by Donald M. Hunten. NASA SP-340. U.S. Government Printing Office, 1974.

MODELS OF THE GIANT PLANETS. M. Podolak and A. G. W. Cameron in Icarus, Vol. 22, No. 2, pages 123–148; June, 1974.

ICARUS, Vol. 24, No. 3; March, 1975.

## THE SMALLER BODIES OF THE SOLAR SYSTEM

MOONS AND PLANETS. William K. Hartmann. Wadsworth Publishing Co., 1972.

THE CHEMISTRY OF THE SOLAR SYSTEM. John S. Lewis in Scientific American, Vol. 230, No. 3, pages 51–65; March, 1974.

PHYSICAL PROPERTIES OF THE NATURAL SATELLITES. David Morrison and Dale P. Cruikshank in Space Science Reviews, Vol. 15, No. 5, pages 641–739; March, 1974.

THE NATURE OF ASTEROIDS. Clark R. Chapman in Scientific American, Vol. 232, No. 1, pages 24–33; January, 1975.

## INTERPLANETARY PARTICLES AND FIELDS

SOLAR WIND AND INTERPLANETARY MAGNETIC FIELD. A. J. Dressler in Reviews of Geophysics, Vol. 5, No. 1, pages 1–41; February, 1967.

INTRODUCTION TO THE SOLAR WIND. John C. Brandt. W. H. Freeman and Company, 1970.

CORONAL EXPANSION AND SOLAR WIND. A. J. Hundhausen. Springer-Verlag, 1972.

SOLAR TERRESTRIAL PHYSICS. Syun-Ichi Akasofu and Sydney Chapman. Oxford University Press, 1972.

SOLAR WIND THREE: PROCEEDINGS OF THE THIRD SOLAR WIND CONFERENCE AT ASILOMAR, PACIFIC GROVE, CALIFORNIA, MARCH 25–29. Edited by C. T. Russell. July, 1974.

Absorption spectra (Venus), 54, 55
Abt, Helmut, 10
Accretion
    Earth, 59, 60
    gravity and, 75
    lunar, 70–71
    outer planets, 107, 110
    planet formation and, 20
    planetesimal, 118–119
Adams, John Couch, 105
Adams, Walter S., 52
Adel, Arthur, 52
Aitekeeva, Z. A., 118
Alfvén, Hannes, 28, 127, 134
Algae, 61, 63, 65
Aluminum, 17, 59, 60
Amalthea, 101, 116, 120
Amino acid synthesis, 63, 120
Ammonia, 5, 10, 17, 60, 63
    bands, 107, 108
Anderson, John D., 97
Angular momentum, 5, 18, 19–20, 23, 39, 70–71
Aphelion, 40, 44
Apollo asteroids, 113–114
    comets and, 121–122
*Apollo* moonprobes, 46, 59, 69–70, 113
    instrumentation, 73
Argon (Mars), 84–85
Asteroid belt, 115–116
Asteroids, 113, 114–116, 117, 120
    craters and, 9
    Mars and, 87
    meteorites and, 116
    Trojan, 115, 116, 118
Astrology, 5
Astronauts, 46, 59, 69–70
Astronomical unit of distance, 5, 6, 18, 19
Astronomy
    early limitations of, 15
    lenses, mirrors and, 5
Astrophysics, 16, 113–114
*Atlas novus coelestis*, 3
Atmosphere, 6
    early Earth, 59, 63, 120
    hydrogen-rich, 10
    Io, 101, 117
    Jupiter, 22, 95, 97–98
    lunar, 69
    Mars, 7, 9, 42, 81–86, 90–91
    Mercury, 42
    Neptune, 22

    outer planets, 106, 107, 108–109
    Titan, 7, 120
    Uranus, 22
    Venus, 49, 52–56
Atoms, solar temperature and, 28
Aurora Borealis, 127
Auroras, solar activity and, 27
Axis
    Jupiter, 6, 96
    Mars, 6, 81, 91

Bands of Jupiter, 96, 97
Barghoorn, Elso S., 61
Barker, Edwin S., 52
Barnard, Edward Emerson, 49–50, 116
Bartels, Julius, 127
Basins of Mercury, 38, 40, 41, 42–43
Beckers, Jacques M., 33
Becquerel, Henri, 28
Benedict, William S., 53
Beryllium and boron, 30
Beta decay, 29
Bethe, H. A., 29
Biermann, Ludwig F., 28, 128
Bigg, E. K., 117
Binder, A. B., 116–117
Biology
    Mars, 81, 84, 91
    planetary, 10
Birkeland, Olaf K., 127
Bode's law, 15, 116
Boron, beryllium and, 30
Boyer, C., 50–51
Brahe, Tycho, 3
Breccia, lunar, 70, 76
Brown, Robert, 117
Buffon, Georges Leclerc de, 15

Callisto, 7, 95, 116, 117
Caloris Basin (Mercury), 38, 40, 42–43, 44, 45
Canals (Mars), 5, 81
Canyon (Mars), 86, 89
Carbon, organic, 63, 65
Carbonate, 56, 85
Carbon dioxide
    Earth, 60, 63
    greenhouse effect and, 7
    Mars, 81, 84, 85, 91
    Venus, 49, 52–53, 56
Carbon monoxide
    Mars, 84
    Venus, 52, 53

Carbon-nitrogen cycle, 29–30
Carleton, Nathaniel P., 51
Carrington, Richard, 27
Cassini, Giovanni Domenico, 96, 98
Cassini's division, 105, 121
Celestial mechanics, 95
Cellular life, 63
Ceres, 116, 120
Chadwick, James, 29
Chandrasekhar, Subrahmanyan, 28
Channels (Mars), 7, 9, 88–89
Chapman, Clark R., 114
Chapman, Sydney, 127
Chemistry
    organic, 10
    planetary formation, 3
Chlorine 37, neutrinos and, 30, 34
Chondrites, 114, 116
Chromosphere, 28, 32
Climate
    Mars, 81, 90
    solar activity and, 27
Clouds
    dry ice, 86
    interstellar, 17, 18, 19
    Jupiter, 98
    Mars, 81, 86
    outer planets, 107
    Venus, 49, 51–53
Coffeen, D. L., 54
Colombo, Giuseppe, 39
Comets, 5, 23, 113, 114, 115
    craters and, 9
    formation, 22, 121
    Jupiter and, 123
    Mars and, 87
    orbits, 22, 121
    tails, 128
Condensation temperatures, 71
Connes, Janine, 52, 53
Connes, Pierre, 52, 53
Continental drift, 64, 65
Convection
    current, 9
    planetary systems, 3, 110
    solar heat, 28
Copernicus, Nicolaus, 3, 37
Cores
    Earth, 59, 60
    gases, 22
    iron, 3, 37, 42, 44, 46, 97
    Jupiter, 95, 97

Cores (*continued*)
  Mars, 84
  outer planets, 107, 109, 110
  Venus, 55
Coriolis effect, 98
Corona, solar
  heat of, 27–28
  magnetic activity and, 32
Coronal hole, 31, 32
Corpuscular radiation, 127–128
Cosmic rays
  galactic, 133
  intensity of, 130
  solar wind ions in, 127
Craters, 3, 9, 87
  lunar, 73, 122, 123
  Mars, 81
  Mercury, 38–41, 46, 114
  planet accretion and, 16
  solar system collision and, 122
  terrestrial, 59, 123
  Venus, 56, 114
Cruikshank, Dale P., 116–117
Crust
  Earth, 59–60, 61, 65
  lunar, 72–75, 114
  Mars, 83, 84
  Mercury, 41, 42
  Venus, 55, 56
Crustal plate (Earth), 61, 65
Crystal fractionation, 72–74

Danielson, Robert E., 106
Davis, D. R., 114
Davis, Raymond, Jr., 31, 34
Days, 6
  Mars, 6, 81
  Venus, 6, 51
Deimos, 113, 114
Dence, Michael R., 123
Density
  internal sun, 28
  Jupiter, 95, 96–97
  Mars, 81–82
  planets, 6
Descartes, René, 15
Deuterium
  proton-proton chain and, 29
  Venus, 56
Diatoms, 65
Diopside, 71
Distribution of mass, solar system, 5, 117, 122
Dole, Stephen H., 10, 11
Dollfus, Audoin, 54, 119
Doppelmayer, Johann Gabriel, 3
Doppler shift, 49
  outer planet rotation and, 106
  Venus, 56
Douglass, A. E., 34
Dry ice clouds, 86
Dunham, Theodore, Jr., 52
Dust
  interstellar, 16–17
  Mars, 81, 89
Dust storms (Mars), 86–87, 90
Dyce, Rolf B., 39

Eagle Nebula, 15
Earth, 3, 6, 59–65
  asteroids and, 114
  atmosphere, 59, 63, 120
  core of, 46
  geologic time, 60–61, 62–63, 65
  magnetopause, 101
  magnetosphere, 101

Mercury and, 42
meteorite and, 114
orbit and sunspot cycle, 33
ozone, 108
properties, 6, 107
Eclipses
  Eros and, 114
  timing by, 106
Eddington, A. S., 28
Einstein, Albert, 28
Electrons
  annihilation, 28–29
  space, 101, 127
  velocity, 133
Electron binding, solar temperature and, 28
Elements
  gravity and, 60
  Jupiter, 95, 99
  Mars, 81, 91
  primordial, 17, 18, 22
  radioactive, 59
  solar, 27, 29
  temperature and, 70
Emden, Jacob Robert, 27
Encke's division, 121
Energy
  conserved, 29
  dispersed, 19
  matter as, 28–29
  neutrino, 29, 31
  primordial thermal, 97
  proton-proton chain, 29–30
  solar, 27–34, 108
Equator (Jupiter), 96
Eros, 114
Erosion (Mars), 81–84, 88–90
Ethane (Saturn), 7
Europa, 95, 116, 117
Evolution
  Earth, 59–61, 63–65
  Jupiter, 95, 97
  lunar interior, 74–75
Exobiology. *See* Biology
*Explorer* spaceprobe, 128, 129, 130

Fabricius, Johannes, 33
Fanale, Fraser P., 117
Ferraro, V. C. A., 127
Flare magnetism, 33
Fluorocarbon, atmosphere and, 9
Foraminifera, 65
Forbush, Scott E., 133
Fossil craters, 123
Fossil records, 61, 65
Fragmentation, gas-cloud, 22
Friedman, Herbert, 28

Galaxies
  formation, 16–17
  solar system and, 5
  spiral, 16–17
Galilean satellites, 10, 95, 100, 101, 116, 117
Galilei, Galileo, 5, 33, 95, 116
Galle, Johann Gottfried, 105
Ganymede, 7, 95, 116, 117
Gas drag, 71
Gas-cloud fragmentation, 22
Gases
  angular momentum and, 19
  eddies of, 18, 20
  electrical conductivity of, 32
  flow in magnetic fields, 31
  interstellar, 16–17, 132
  ionized, 127
  Jupiter atmosphere, 98

Mars, 84, 85
outer planets, 106–108, 109, 110
planet formation and, 17–18, 22, 60
solar interior, 29
solar nebula, 17
Gault, Donald E., 41, 44
Gehrels, Tom, 54, 116
Geologic clock, 60
Geologic features, 81, 82
Geologic process of planets, 16
Geology and solar theory, 27
Geomagnetic field cycle, 128
Geophysical station, lunar, 73
Gierasch, Peter J., 91
Glaciation, 64–65
Goguen, Jay D., 7
Goldreich, Peter, 20, 39, 122
Goldreich-Ward instability mechanism, 20–21
Goldstein, Richard M., 49
Gorges (Mars), 81
Gravitational field, 20–21
  Jupiter, 95–97
  planetesimal, 123
Gravity, 3, 6
  accretion and, 75
  Earth, 59, 60
  energy of, 82
  instability mechanism, 20–21
  Mars, 81, 84
  Mercury, 41
  planet density and, 16
  solar nebula formation and, 18, 19
  Venus, 51
Great Red Spot, 5, 96, 98–99
Greeley, Ronald, 90
Greenhouse effect, 7, 55, 60, 91
Guinot, Bernard, 50–51

Hansen, J. E., 54
Hapke, Bruce W., 44
*Hawkeye*, 129
Hawkins, Gerald S., 122
Heat
  Jupiter, internal, 95, 97, 98
  Mercury, emission of, 38, 39
  solar flow, 28
  young star dispersion of, 19
Heliopause, 132
*Helios*, 131
Heliosphere, 133
Helium, 5, 17, 29–30
  hydrogen into, 29–30
  Jupiter and, 22, 95, 97
  Mercury and, 42
  outer planets and, 22, 107
  solar, 27, 29
Herbig, George H., 123
Herschel, William, 105
Hidalgo, 115
History, theories of, 7
Hooke, Robert, 5
Houck, James R., 84
Hubbard, William B., 97, 109
Humason, Milton, 105
Humidity (Venus), 52
Hurricanes, 87, 99–100
Huygens, Christiaan, 5, 10
Hydrocarbons (Vesta), 7
Hydrogen, 3, 5, 17, 63
  absorption, 107, 108
  galaxy formation and, 16
  helium and, 29–30, 95
  Jupiter, 95, 97
  Mars, 84
  outer planets and, 22

solar, 23, 27, 29
Titan, 120
Venus, 56

Iapetus, 10
trailing hemisphere of, 120–121
Icarus, 114
Ice, 10
Mars, 9, 81, 84
planet formation and, 17–18, 22
rings of Saturn, 109
Titan, 120
Ice Age, 64–65
solar energy and, 31
Igneous processes, 71
Igneous rock, 59–60, 61
lunar, 69–70, 71, 72, 73
Venus, 55, 56
Imbrium Basin, 38, 43, 44
Infrared absorption band (Venus), 52
Infrared spectography (Mars), 86
Instrumentation, *Apollo*, 73
Interstellar matter, 16–17, 20–23, 123, 132
Io, 10, 95, 100–101, 109, 116–117
Ion gun, 131
Ionosphere, 100–101
Ions in space, 127
Iron planet core, 3, 37, 42, 44, 46, 97
Isaacman, Richard, 10, 11
Isotopes, 16, 82, 83, 84, 85
rare Venus, 52
Isotopic timers, 76–77
Iversen, James D., 90

Janssen, Michael A., 55
Janus, 119
Jupiter, 5, 6, 63, 87, 93–101, 106–110
atmosphere, 22, 95, 97–98
bands, 96, 97
clouds, 98
comets and, 22, 121, 123
cosmogony, 95, 97
elements, 17, 18
exploration, 4
jet streams, 99–100
magnetic field, 95, 97, 100–101, 117
magnetosphere, 7, 100–101, 127
moons, 5, 7, 95, 101, 113, 116, 117
planetesimals and, 123
satellites, 117–118
structure, 95–97
temperature, 97
Trojan asteroids and, 115, 116
wind, 96, 99–100
year, 5, 6

Kant, Immanuel, 15
Kaula, William M., 123
Keeler, James, 121
Kepler, Johannes, 5
Kinetic energy, lunar temperature and, 75
Kinks, magnetic field, 132
Kirkwood gaps, 116
Kliore, Arvydas J., 117
Knowledge, expansion of, 3
Kowal, Charles T., 116, 118
KREEP norite, 70, 71, 76
Kuiper, Gerard P., 39, 107, 113, 119, 121
Kuiper Crater, 38, 39, 40

Lagrange, Joseph Louis, 116
Laplace, Pierre Simon de, 15
Larson, Stephen M., 105
Leverrier, Urbain, 105
Levy, Saul, 10

Lewis, John S., 54, 108
Life, environment for, 63, 120
Mars, 7, 10, 81, 84, 90–91
plant, 3, 61, 63–65
Light-year, 5
Lindemann, F. A., 127
Lithosphere, 61
Little Red Spot, 96
Lobate scarps (Mercury), 41–44
Loops, solar, 27
Low, Frank J., 122
Lowell, Percival, 105
Lowman, Paul D., Jr., 113
*Luna* moonprobes, 73, 129
Lunar rock, 69–77
Lunar theory, 70–71. *See also* Moon (Earth)
Lyot, Bernard, 54
Lyttleton, R. A., 121

McCord, Thomas B., 114
McElroy, Michael B., 9, 85
Magnetic fields, 9
interplanetary, 101
Jupiter, 95, 97, 100–101, 117
Mercury, 38, 42, 45, 46
particles and, 127–134
planet dynamics and, 16
radial component, 131
solar prominences and, 32
umbra, 33
Venus, 56
Magnetic pole reversal, 64
Magnetic storms, 27, 28, 127
Magnetism, flare, 33
Magnetogram, sunspot groups, 30
Magnetopause, 101
Magnetosphere, 7, 100–101, 127
Mantle
Earth, 59
Luna, 73
Mars, 84, 85
Titan, 109, 113
Venus, 55
*Mariner* spaceprobes, 7, 9, 37–46, 50, 51, 54, 55, 81, 86, 113, 114
Mars, 6, 80–91
atmosphere, 7, 9, 42
biology, 81, 84, 91
canals, 5, 81
channels, 7, 9, 88–89
clouds, 81, 86
cross-section, 84
exploration, 4
maps, 81
Mercury and, 37, 40, 41, 42, 45
moons, 83, 113, 114
orbit, 4–5
tectonic movement, 42, 87–88
temperature, 81, 82, 84, 86, 90–91
volcanism, 9, 81–84, 87–88
winds, 9–10, 81–84, 86, 87
year, 6, 81
Marsden, Brian G., 114
Mass, 6
distribution of, 5, 117, 122
energy and, 28
Jupiter, 95
Mars, 82
mean density and, 106
Mercury, 37
solar nebula, 18–19, 20, 21
Matter, energy as, 28–29
Maunder, E. W., 34
Maxwell, James Clerk, 109, 121

Mercury, 6, 36–46, 114, 120
asteroids and, 115
atmosphere, 42
basins, 38, 40–43
exploration, 4
orbit, 4–5, 41, 44
temperature, 38–40
Metazoan, 63, 65
Metal, alkali, 59
Meteorides
Mars and, 87
near miss, 114
Venus, 56
*See also* Meteorites
Meteorites
Earth orbit and, 114
materials, 70
research on, 15–16
*See also* Meteorides
Methane, 5, 60, 63, 64
ices, 10
Mars, 84
outer planets, 106–107, 108, 109
Titan, 120
Milky Way, 5
Mineral compound, phase diagrams, 71–72
Minerals
Mars, 81
Mercury, 44
Minton, R. B., 116
Models
Earth, 82–83
Jupiter, 95, 97
lunar accretion, 70, 82–83
massive-nebula, 19
nebula evolution, 20
Molecules, complex, 3
Moon (Earth), 63, 64, 69–77, 101
breccia, 70, 76
craters, 5
exploration, 3, 113
formation, 114
Imbrium Basin, 38, 43, 44
lunar theory, 70–71
Mercury and, 37, 40, 41, 45
rock, 69–77
temperature, 75
tidal forces, rotation, 51
Moons
Jupiter, 5, 7, 95, 101, 113, 116, 117
Mars, 83, 113, 114
outer solar system, 10, 106
Saturn, 106, 113, 119, 120
*See also* Satellites
Moroz, V. I., 52, 55
Motion, planetary, 2–3
Mountain building, 9, 60
Mrkos, 132

Natural selection, 3
Nebulae
gaseous emission, 15
interstellar matter and, 16–17, 20–23, 123, 132
stellar system, 118, 119, 122
*See also* Solar nebulae
Nebular hypothesis, 3, 15, 19, 20
Nelson, David, F., 127
Neon, 17
Neptune, 6, 105–108, 110
atmosphere, 22
elements, 17, 18
Pluto and, 4–5
rotation, 10
satellites, 121

Ness, Norman F., 42
Neutrinos
    argon 37 and, 30, 34
    detection, 30–31, 34
    energy in, 29, 31
Neutron particle, 29
Nitrogen (Mars), 85
Norite. See KREEP norite
North pole (Jupiter), 95
Nuclear fusion, plasma physics and, 127
Nuclear transformation, 28–30
Nuclei, solar collision and, 29

Occultation. See Eclipses
Ocean (Earth), 59, 60, 61, 65
O'Leary, Brian, 114
Oort, Jan, 22
Oort cloud, 22–23
Opik, Ernst, 123
Orbits
    asteroid, 115
    comet, 22, 121
    control, spacecraft, 131
    Jupiter, 98
    Mercury, 4–5, 41, 44
    physics of, 70–71
    planetary, 4–5, 6
    satellites, 118
Organic chemistry, 3
Organic compound, early Earth, 63
Organisms
    early Earth, 3, 61–65
    Mars, 90
    See also Biology; Life
Outer planets, 105–110. See also Jupiter;
        Neptune; Pluto; Saturn; Uranus
Outgassing, 85, 91
Oxygen
    and life, 63, 64, 65
    lunar isotopic, 69
    Mars, 81, 84, 85
    Venus enigma, 53
Oxygen VI, 27
Ozone, 53, 63
    fluorocarbon and, 9
    Mars, 84, 85

Paleobiologic record, 64–65
Paleoclimatology, 9
Paleomagnetic activity, 64
Pallas, 116, 120
Pangaea, 65
Parker, E. N., 129
Particles
    ionized, 117
    lunar accretion, 70–71
    orbit, 23, 132
    space, 127–134
Pauli, Wolfgang, 29
Peale, Stanton J., 39
Penumbra, magnetic field, 33
Perihelion (Mercury), 40, 44
Perri, Fausto, 21, 23
Perspectives, cosmic, 4–5
Pettengill, Gordon H., 39
Pettijohn, Francis J., 61
Phobos, 113, 114
Phoebe, 120
Photochemistry (Mars), 85
Photodissociation (Mars), 85
Photography, astronomical, 7
Photons, particles and, 23
Photosphere, 28
Physics, space, 16, 113–114, 127

Pickering, W. H., 105
Pilcher, Carl, 121
Pine, Milton R., 19
Pioneer spaceprobe, 7, 95–99, 101, 105, 109,
        117, 129–130, 131, 132
Plains
    Mars, 82–83
    Mercury, 41, 43
Planets, 5, 6
    asteroid and, 115–116
    elements, 17
    exploration, 2–10
    forming, 9, 17, 20, 21, 22, 67–70, 82
    gravitation, 16
    magnetic field, 16
    mass, 17–18
    properties, 6–7, 107–109
    satellites and, 10, 22, 113, 114, 117–118,
        119, 120
    See also Cores
Planetary motion, 2–3
Planetary system, alternative, 11
Planetesimals, 116, 117, 118–119, 121
    Jupiter and, 123
    solar system bombardment and,
        114, 122–123
Planetology, 7, 9
Plant life, 3, 61, 63–65
Plasma
    electric current in, 129
    energy exchange in, 131
    physics, 127–128
Plate tectonics, 3, 61, 64, 65. See also
        Tectonics
Pluto, 6, 105–106
    orbit, 4–5, 121
    rotation, 10
Podolak, Morris, 18
Polarization, light, 54
Poles
    Jupiter, 95, 96, 98
    Mars, 9, 81
Pollack, James B., 9, 120, 121
Potassium, 59, 60
Poynting-Robertson effect, 120–121, 122
Pressure, atmospheric
    Mars, 6, 7, 9, 84
    Venus, 6, 52
Pressure, interior, (Jupiter), 97
Processes, chemical interaction, 16
Properties, planets, 6–7, 107–109
Protogalaxies. See Galaxies
Proton-proton chain, 32, 34
    deuterium from, 29
    energy of, 28
Protons
    annihilation, 28–29
    energy, 133
Ptolemaic tradition, 3
Pyramids (Mars), 9–10

Radar exploration, 7, 9, 49
Radar map (Venus), 55–56
Radiation
    heat energy and, 19
    particles and, 23
Radiation belts
    Jupiter, 101
    particles, 127
Radioactive isotopes, 16, 82, 83, 84, 85
Radioactivity, 3
    Earth, 59, 60
    lunar soil, 74–75
    rock dating and, 60–61
    solar energy and, 28–29

Venus, 56
Radio-astronomy, 127
Radio-emission
    Callisto, 7
    Io, 101
    Jupiter, 95, 96, 100–101
    Saturn, 7
Radio-reflectivity map (Venus), 49
Rainfall (Mars), 89
Ramsey, William, 27
Rayleigh scattering, 95
Richardson, Robert S., 49
Rings of Saturn, 5, 9, 105, 109, 121
Rock
    Earth, 59–61, 64, 65
    lunar, 69–77
    planet formation and, 17–18, 22
Rogers, Ernest H., 127
Ross, Frank E., 50–51
Rotation, 6
    Jupiter, 95–96
    magnetic field and, 56
    retrograde, 49–51
Rutherford, Ernest, 29

Sagan, Carl, 90, 91
Sand dunes, 9
Sapping on Mars, 88, 89
Satellite system, planets, 10, 22
Satellites
    Jupiter, 116, 117–118
    planets, 6, 113, 114, 116, 117–118, 119,
        120, 121
    See also Moons
Saturn, 6, 105–110
    atmosphere, 22
    elements, 17, 18
    moon, 106, 113
    radio bursts, 7
    rings, 5, 9, 105, 109, 121
    satellites, 119, 120
Savage, Blair, 106
Scarp, lobate, 41, 42, 43, 44
Schorn, Ronald A., 52
Schröter, E. H., 33
Schultz, Peter, 44
Schwabe, Heinrich, 34
Schwarzschild, Karl, 28
Scintillation, 131
Seasons (Mars), 81
Sedimentary rock, 60, 61, 63, 65
Sedimentation (Mars), 81, 90
Seismic studies, lunar, 73
September solar system, 4–5
Shapiro, Irwin I., 39
Shoemaker, Eugene M., 114, 122
Silicate, 60
Silicon, 59, 60
Silt (Mars), 88, 90
Singer, S. Fred, 121
Sinton, William M., 52, 55
Skylab, 27, 28, 29, 31, 32
Slipher, V. M., 52
Smog, outer planets, 108
Sodium (Io), 10, 117
Soil
    lunar, 73
    Mars, 84
Solar activity
    Earth's magnetic activity and, 32
    terrestrial environment and, 34
Solar electric propulsion, 131
Solar energy, 28–34, 108
Solar flares, 27–29
    particles emitted by, 133–134

Solar heat, 28–30
Solar magnetic storms, 27, 28, 101
Solar nebulae, 23, 110
   accretion time, 19, 20–21, 22
   cooling time, 19–20
   Earth and, 59, 60
   formation, 3, 15, 16, 17
   gases of, 16
   meteorites and, 116
   properties, 17, 18–19, 20, 21
   radiation, 19
   *See also* Nebulae
Solar radiation, Mercury and, 40–41
Solar surface, spacecraft and, 27
Solar system components, 122–123
Solar wind, 23, 127–134
   coronal hole and, 32
   Earth and, 101
   Jupiter and, 101
   magnetic activity and, 28
   Mercury and, 42, 46
   speed, 127
   Venus and, 56
Soter, Steven, 44, 120–121
Space
   astronomical unit distance, 5, 6, 18, 19
   electrons in, 101, 127
   exploration, 4
   physics, 127
Spacecraft, 3–4
   interplanetary, 7–9
   television cameras, 86
   trajectory, 33, 131
Space-probe program, 16
Spectrography
   Jupiter, 98
   Mars, 81
   Venus, 51, 54–55
Spin-orbit coupling (Mercury), 39, 40, 44
Spiral clock, evolution and, 63
Spiral galaxy, 17
Stars
   formation, 16–17
   nebula of, 122
   T Tauri stage, 19
Stellar bodies, gravity and, 20
Stellar system hypothesis, 122–123
Stewart, John, 19
Störmer, Carl, 127
*Stratoscope II*, 106
Stratosphere, outer planets, 108–109
Strom, Robert G., 41–42
Stromatolite, 64
Strong, John, 52, 55
Subatomic-particle concept, 28
Sulfuric acid, atmospheric, 49, 54–55
Sulfuric acid hypothesis (Venus), 54–55
Sun
   age, 27
   angular momentum, 19
   asteroids and, 115
   comets and, 121
   cross section, 28
   life expectancy, 30
   mass, 18–19
   material opacity, 28
   Mercury and, 37, 44
   solar wind, 23, 127–134
   T Tauri phase, 23
   theories, 15, 27
   *See also* entries under Solar
Sunlight, 23
   Mercury and, 37

Sunspots, 27
   behavior, 33–34
   cycles, 34, 130
   duration, 28
   early records, 33–34
   groups, 30
   magnetic activity, 32–33
Supernova, 18
   cosmic ray source, 134
   explosion, 17
*Surveyor 7*, 73
Synchrotron radiation (Jupiter), 100
Syrtis Major, 5

Taylor column, 99
Tectonics
   Earth, 61, 65
   Mars, 42, 87–88
   Mercury, 41, 42, 44
   Moon, 41
   *See also* Plate tectonics
Telescopes, 32
Temperature, 6
   Jupiter, 97
   Mars, 81, 82, 84, 86, 90–91
   Mercury, 38–39, 40
   solar, 27, 28
   umbra, 32
   Venus, 49, 52–55
Terrain
   lunar, 69
   Mars, 87, 90
Terrestrial environment and solar activity, 34
Theories
   lunar, 70–71
   planet origin, 15
   solar nebula mass, 19
   sun, 15, 27
Thermal convection currents, outer planets, 109, 110
Thermal emissions, outer planets, 106, 108–109
Thermal radiation, 7
Thermodynamic law, second, 28
Thermonuclear reaction, solar, 28
Thorium, 59
Tidal force (Venus), 51
Tides (Earth), 64
Time-space, 5
Titan, 107–110
   atmosphere, 7, 119–120
   hydrogen, 10
   mantle, 109, 113
Titanium (Mercury), 44
Tomasko, Martin, 106
Tombaugh, Clyde, 105
Toon, Owen B., 9, 91
Topography
   Mars, 81–84
   Mercury, 41, 43
Toro, 114
Trafton, Laurence M., 107, 108
Transonic meteorology, 9
Trask, Newell J., Jr., 45
Traub, Wesley A., 51
Tree-ring dating, sunspot cycle and, 34
Triton, 120, 121
Trojan asteroids, 116, 118
   Jupiter and, 115, 116
T Tauri
   stage, 19, 23
   wind, 23
Tyler, Stanley A., 61

Ulrichs, J., 44
Ultraviolet photography, 49, 50, 53
Ultraviolet radiation, 63
   ozone and, 85
Umbra temperature, 32
Urania, 2–3
Uranium, 59
Uranus, 6, 105–110
   atmosphere, 22
   elements, 17, 18
   rotation, 7, 10
   satellites, 121

Vacuum, space, 127
Vaiana, Giuseppe, 28
Van Allen belts, 128–129
*Venera* spaceprobes, 51, 52, 53–54, 56
Venus, 4–5, 6, 7, 9, 48–56
   atmosphere, 49, 52, 56
   clouds, 49, 51–53
   craters, 56, 114
   environment, 7, 60
   exploration, 3–4
   Galileo and, 5
   magnetic field, 42
   Mercury and, 44, 45, 46
   properties, 6
   surface radar of, 9
   temperature, 49, 52–55
   water, 9, 49, 52–55, 56
   weather, 55
   wind, 51, 56
Veverka, Joseph F., 119
*Viking* spaceprobe, 91
Volcanism
   atmosphere and, 9
   Earth, 60–61, 72
   lunar, 69, 73
   Mars, 9, 81–84, 85, 87–88
   Mercury, 43, 44
   planets, 16
Volume
   Jupiter, 95
   planets, 6

Wallace, L. W., 108
Ward, William R., 20, 91, 122
Water, 3, 60, 63, 65
   greenhouse effect and, 7
   ices, 10
   Jupiter vapor, 7
   Mars, 81, 82, 84–87, 88–89, 91
   stellar bodies and, 22
   Venus and, 9, 49, 52–55, 56
Weidenschilling, S. J., 123
Wetherill, George W., 45, 123
Whipple, Fred L., 22, 114
Wildt, Rupert, 109
Williams, James, 114
Wind, 89–90
   Jupiter, 96, 99–100
   Mars, 9–10, 81–84, 86, 87
   Venus, 51, 56
   *See also* Solar wind
Worden, Alfred M., 69

X-rays
   lunar radiation, 73
   solar, 28

Year
   Jupiter, 5, 6
   Mars, 6, 81